Song of Hope

Song of Hope

The Green Revolution in a Panjab Village

Murray J. Leaf

Rutgers University Press, New Brunswick, New Jersey

Photographs by Murray J. Leaf

Library of Congress Cataloging in Publication Data

Leaf, Murray J.
Song of hope.

Bibliography: p.
Includes index.
1. Green Revolution—India—Punjab—Case studies.
2. Agriculture—Economic aspects—India—Punjab—Case studies. 3. Villages—India—Punjab—Case studies.
4. Sikhs—India—Punjab—Case studies. I. Title.
HD2075.P8L42 1984 338.1′0954′552 83–11202
ISBN 0–8135–1025–2

Frontispiece:
The wedding season 1965; a bridegroom's dance of joy.

Contents

Figures

Tables

Preface

Even though the Green Revolution has brought great social and economic success to the Indian Panjab and much more progress is still possible, in Western scholarly circles there is doubt that the revolution ever existed, and in some Indian political circles the belief is developing that the revolution is over and attention is shifting to dividing its fruits.

To convey what has happened in its full breadth poses a basic challenge to anthropology, which, alone among the social sciences, has declared itself to have assumed the task of translating the experience of one human community so that it can be understood by members of others, without reductionism and without setting aside the beliefs, values, and convictions of those being described.

The civil disruptions that have been increasing in intensity in Panjab since about 1980, although often represented in the Indian national media as a conflict between secularism and communalism, in fact represent quite different assessments of the Green Revolution and the best way to spread or share its benefits, and they dramatically show how urgent it is to gain a dispassionate and factual understanding of what has happened. In a different context, it is equally urgent to show, by a concrete example, what anthropology has to contribute to the analysis of what may seem a narrowly technical rather than a globally "cultural" phenomenon and, conversely, what anthropology itself can gain from being so applied in the inevitable and necessary clash with other well-articulated and vigorously defended scholarly disciplines.

When I returned to Panjab in the spring of 1978, it was for four reasons, none of which anticipated the present study. First, I had been away too long. Second, I had hopes of making arrangements for a study of a pattern of migration that had been important at the time of my first study, from 1964 to 1966, and that might have enabled me to extend my previous work to a larger area and some new problems. Third, I had read much in the scholarly literature and in news media about the Green Revolution, but could make no consistent pattern of it. There was so much disagreement that it seemed arbitrary to trust any one opinion. And last, I received an invitation from Manohar Singh Gill, then newly appointed Principal Secretary to the Chief Minister of Panjab, who said the time was good for me to return and see what had happened. During my first study, Gill had been District Commissioner of Ambala district, which was the administrative division that contained the village I concentrated on. We had talked occasionally and later we corresponded, and I had always been impressed with his grasp of my own work, with his sense of the possible role of social analysis in policy formulation, and with the precise aptness of his advice and observations. I was confident that if he said the time was right, it was.

My intent was to follow the same general pattern of work I had the first time: to have a place in Chandigarh and another in the village, and to move between them as the work required. In addition to possessing such amenities as running water and electricity, Chandigarh is a very pleasant city and there were many friends I was anxious to see again. It also happens that as an educational, transport, communication, and administrative center, Chandigarh was the best place to arrange for whatever records and permissions I might require, obtain technical help and advice, buy books and the like, and get some restaurant meals to minimize my impositions on the villagers, who I knew would never allow me to pay for food or lodging. I would travel to the village for work periods of one to a few days, generally with specific questions to pursue, and go back to Chandigarh to write up the results and get background information as need arose. Since it was the hot season, there would be little work in the fields to watch, but it would be a good time to talk.

My plane touched down in New Delhi (six hours late) on the morning of June 9, just as the sky was brightening above a haze thick with the smells of diesel and coal smoke, native tobacco, wet manure, dry earth, and jasmine. The city was in fact suffering an exceptional heat wave, and by that afternoon I had determined that most of the people I planned to visit were off in the mountains. But late that night it rained, and the next morning was cloudy, with the air fresh and clean. Summer had ended—as suddenly as it can only in India. The paper at breakfast said it was just a "pre-monsoon" shower, but I wanted to avoid any possible chance of being cut off from the village (since it lay north of Delhi, the rains would start there later). So I spent the day hastily making a few necessary purchases and travel arrangements and the next morning took the first bus north (forgetting what ordinary-class buses can do to knees, but enjoying the opportunity to reenact my first such passage). We passed back into the hot season about eighty kilometers up the road.

On June 12 I was in the village. The magnitude of the Green Revolution and the need for describing it became clear that same day.

In addition to the crucial debt I owe M. S. Gill for his support and advice in all aspects of the project, it is a pleasure to acknowledge the hospitality and guidance on many points of law and legal history of Sr. S. S. Sodhi, then Registrar of the Panjab High Court. I am also pleased to thank Mr. Gurcharan Singh Kalcatt for his comments on the last revision of the manuscript of this study.

Last of all and most important, I must once again thank the villagers of Shahidpur, especially Naranjan Singh and his family, and Surindar and Netar Singh, their father, Indar Singh, and his family. After I returned to the United States, Surindar took it upon himself to continue to answer questions as they arose from the analysis and to read and respond to a full manuscript of this account. This project was undertaken in response to their encouragement and could not have been completed without their help. It is dedicated to their hopes.

With only a short summer to cover so much, the usual author's assumption of responsibility for any errors or omissions takes on a special poignancy. I especially regret that I could not follow the procedure of the first study and discuss a completed draft of this work with the villagers face to face. With more money and time, the job would surely have been better done.

Census and landholding information for 1978 was collated and analyzed, and the 1965 data reanalyzed and compared with it, using the computer facilities of the University of Texas at Dallas. Primary support was provided by a small grant from the National Institute of Mental Health (1 RO3 MH33436-01), with additional support from the University. Programming was done by Mr. Paul Chu. Mrs. Evelyn Stutts, Mrs. Ruth Cole, and Mrs. Annette Corken helped prepare the manuscript.

My wife, Michelina, provided help and support in the initial field study, in editing the manuscript, and in almost all other phases of the work.

The kinship chart printed here as Figure 6.1 first appeared in *American Anthropologist* (Leaf 1971). The two supplementary charts shown here as Figures 6.2 and 6.3 appeared originally in my *Information and Behavior in a Sikh Village* (Leaf 1972). The data presented in Tables 3.1 through 3.4, 4.1, 4.3, 5.4, and 5.6 were first given in Tables 1 through 7, 9, and 11 in the winter 1983 issue of *Economic Development and Cultural Change* (Leaf 1983). Most of the information in Table 5.1 first appeared in *Pacific Affairs Quarterly* in 1981 (Leaf 1981).

Note on Orthography

The consonants used in transliterating Panjabi terms have the same values as in Italian and Spanish, except that capitalization indicates stops of the retroflex series (made by curling the tip of the tongue back slightly to touch the back edge of the alveolar ridge). The capitalized "R" is a retroflex voiced continuant.

The values of the vowels are as follows:
a as in but
ā as in pot
i as in bit
ī as in beet
e as in bet
o as in bought
u as in foot
ū as in boot
ai as in bite
au as in boat

North Indian scripts, including both the Urdu cursive and Gurumukhi in which Panjabi is most commonly written, do not have a distinction between capital and lower-case letters, and there are no writing conventions analogous to those concerning capitalization of proper nouns in English. Accordingly, transliterated proper nouns other than personal names (which are not italicized) are not capitalized here, unless they appear at the beginning of a sentence.

There are now many English words used in Panjabi, and there is also a distinctive form of English in use in north India that has numerous naturalized words from Panjabi, Hindostani, Persian, and other South Asian languages. English words used in Panjabi and naturalized words commonly used in Indian English have been treated here as ordinary English words.

English translations have been used as the names of castes except where there are no convenient counterparts. Words used as caste names have been capitalized. The names of actual occupations have not been. Thus there are many members of Leatherworker caste, but in 1978 no one in the village earned money as a leatherworker.

Song of Hope

CHAPTER 1

Introduction

The Green Revolution has been a focus of national and international concern for two decades. It has involved millions of acres of land and hundreds of millions of people. Yet opinions on its causes, nature, and effects range from the view that it promises "food and jobs for all" (Sen 1975), through the view that its hidden costs outweigh its gains or might do so in the future (Billings and Singh 1971; Falcon 1971), all the way to the view that it has made a bad situation worse—increasing economic inequality and political instability (Frankel 1971).

The disagreements present a basic challenge to social theory. Some scholars have suggested that they cannot be resolved in principle within the framework of a purely factual and analytic social science. This suggests in turn that social science has to be grounded in political ideology—a view that leads ultimately to the substitution of violence for reason. In my own view, the disputes are not by any means a reflection of defects in social analysis as such, but only in some of its specific methods.

The Green Revolution has not been one thing but many things. Yet the method too often chosen to study it has been to select one facet or relationship at a time, often from several distinct areas—such as the relationship between inflation and production over time or the displacement of labor by mechanization. This kind of focus gives an air of precision, but it often ignores the point that inflation can have very different significance depending on exactly which goods are priced and their relative changes in value within each region, just as labor displacement may mean very different things in different places depending on exactly who is displaced and what other opportunities they have. Each thing must be taken in its own actual context, and this simply cannot be done with data aggregated at national or even regional levels. The remedy is simple, though it may at first seem retrogressive. It is to allow the interpretation of seemingly broad and general phenomena, such as labor rates and price rises, to wait until they can be seen in the context of detailed studies of particular communities—even just one community. When this is done, the realities of the revolution come to life and the ambiguities of the isolated measures disappear. We can obtain a definitive understanding of the underlying motivations of those who responded to the new opportunities, an understanding of the relationship between the decision-making processes of villagers and the governmental and other policies they responded to. We can see the reasons for what happened and beneath the reasons we can see the reasonableness, and from this we can understand why certain policies or incentives worked and were accepted and why others did not and were not. Of course each village is unique and each region is unique, but these underlying processes are not. They are what we are mainly concerned with understanding, and they are where leading issues in social science come together with basic questions of policy and ethics.

My first field work in the Panjab village I call Shahidpur began in October 1964 and ended in February 1966. In 1978 I returned to the village for the summer. It happens that 1965 and 1978 come closer than any other years to marking the onset and completion of the Green Revolution in Panjab. I did not foresee this in 1964, but the data collected then still provide an excellent basis for examining the revolution in retrospect. The story has many threads and therefore many beginnings. The simplest way to start is with the way this project has grown from my earlier relations with the village.

M. N. Srinivas, speaking of the village he lived in and studied in Mysore, remarks that "looking back . . . I cannot help thinking that three individuals . . . contributed significantly to my understanding of village life and culture" (1976:53). There is a reason for this which goes far beyond the extraordinary way anthropological field work requires the analyst to depend upon those whom he studies and beyond the fact that there are impressive individuals in every community. It has to do with the nature of human communities themselves.

An earlier generation of anthropologists often spoke of the native "perception" of his own society as if it were something the analyst could ignore or take into account at his pleasure while describing the community "itself." But a community is not something the natives merely perceive, something which an anthropologist or other outsider might see for himself apart from them. It is something they live and it is something that lives only through them. What one learns is not "the" village, but always someone's village—hopefully someone whose perspective will be cogent and meaningful to most others. Ultimately, one puts together many such personal perspectives, making the result one's own just as villagers themselves do. Exactly how these perspectives are obtained and assembled and how the result is explained depend on one's theoretical perspective. Mine will be described in the next chapter. For the moment, it is only important to say that precisely because one can only come to know a village through those who live in it, selecting a village to work in presents the gravest perplexities.

Social scientists generally like to claim that what they describe is "typical" or representative of *something* larger, somehow. My intent in 1964 was to study a Panjabi Sikh village, but since I had not yet done the study, I had no firm idea of what characteristics were to be expected in villages of this type. Census tabulations available in the United States had been uninformative for precisely the same reason that the aggregated data used to describe the Green Revolution have led to so much confusion. I knew of no better data that might be available in India. My plan was to cross the region a few times to get a feel for the parameters of variation and then make a selection on the most logical basis that emerged. But as my clattering ordinary-class bus followed the Grand Trunk Road north out of Delhi and into the rural countryside for the first time in early June, the difficulties began to make themselves clear. The plain was dead level and a patchwork of cultivated fields seemed to spread in all directions forever, without break, boundary, or a sign or mark giving the name or limit of any village or town. Some fields were dry, others had green crops. Some were large, some small. Here and there, sometimes in groups, sometimes alone, were oddly trimmed trees with thick trunks and branches cut off short. There were little isolated buildings, larger clusters of buildings, many open areas, ditches, ponds, roads, cattle, and canals. Alongside the roadway there were odd little shops, small sheds more like large boxes. There were a few apparent factories. There were some large towns with what appeared to be suburbs. There were isolated brick kilns. There were small groups of people who must have belonged somewhere sitting under isolated trees. There were white-painted shrines, ancient brick wells, and here and there what looked like a family on a bicycle going somewhere along the road, although no place might be in sight.

Which of these were villages or village related? Where did one village leave off and another begin? Were there nonvillages among villages? There was no hope of finding a typical village if I couldn't identify even one village to start with, and at this level of raw detail it was highly doubtful that I could see any general patterns in weeks or even months of travel. And there were practical problems in addition to the logical ones. The ride was hot, dusty, and uncomfortable. I didn't know the bus schedules or routes, the conventions of travel, or where or how to find accommodations. I didn't know where or how to get a meal, or even what a meal was. Instruction in Panjabi had not been available in the United States, and all I had was a bookish but rudimentary Hindi, which in my first few weeks in New Delhi had already shown itself to produce more confusion than information. Because this was still soon after the Indo-Chinese war in Tibet, road maps were classified documents and could not be obtained by any means. Not knowing this, I had not brought any with me and had not bothered to memorize the major towns and routes of the area. Buses seemed to go from town to town. How did one get them to stop at a village? What was the word for "village" anyway? It might be the Hindi term I had learned, but I could not be sure. Even if I got to a village, what would I say? Who should I talk to? Again, I would have to know much more even to begin.

My first stop was to be Chandigarh, the state capital. It turned out to be a spacious, modern city with wide streets and the smell of flowers always in the air. It is laid out at the foot of the Siwalik hills that lie in turn along the foot of the Himalayas, at the extreme eastern edge of the Panjab plain. The only Western-style hotel was the Mountview, whose rooms shared a common open veranda in front facing west toward the plains and had private balconies in back facing east to the mountains and to Le Corbusier's dramatic capitol complex. The view from my balcony on the second floor, over its trees and broad lawn, took in the administrative buildings, the Siwaliks behind them, and the Himalayas rising beyond them both, and was striking even in this parched time. June is in the middle of the season called, simply, "hot." The sky showed no cloud; there was no wind but a hot *luh* that blew steadily at midday. The sun did not set in any familiar way; rather each afternoon, on its way to the horizon, it seemed to melt into a pall of redhot dust. I still remember, on about my third day (after recovering from a feverish sore throat that began on the ride up), the sinking feeling I had coming out the front door of my room and squinting into such a sunset. Out in the forbidding dusk beyond the whitewashed buildings were ten thousand villages. One of them was the right one, but how would I find it?

Fortunately others understood the problem. R. L. Anand had been doing field studies in Panjab since the early part of the century, in the Panjab Board of Economic Enquiry. In 1964 he was Director of Census Operations for the state. I had read some of his work while still in Chicago. I found that the census office was only a block away and went to see him the next day. He advised me that the 1960 census was just then being tabulated, under the supervision of Professor G. S. Gosal at Panjab University, in Chandigarh. Anand suggested I see him. Professor Gosal in his turn very graciously suggested that I use the raw tallies of the census results for the eighteen districts for a preliminary survey. Districts were units small enough to show truly local relationships among variables, although they were numerous enough to treat statistically. The question was whether I could analyze the returns in such a way as to find relationships that reflected the outlook of those censused rather than simply the imposition of my own arbitrary categories.

The kindness of Anand and Gosal and the detail of the available data were entirely unplanned for. Yet if there was a chance to identify a typical village in a clear statistical sense based on fresh and comprehensive data, the possibility could not be rejected, and the time it would take would also allow me to get help with Panjabi. I bought a small slide rule and began. Since the proportion of Sikhs in villages in the state varied greatly and the average village had less than 50 percent Sikhs, no simple state averages would show what to expect in a village that was predominantly Sikh. It was necessary to see what correlated with the proportion of Sikhs in the rural areas of the state and define typicality by this pattern of associations. This was rather complicated. Later, the Department of Economics of Panjab University kindly allowed me to use a Facit calculator. It took three months to set up and go through the sixty or seventy most-promising comparisons, but eventually a quite clear pattern emerged associating adherence to the Sikh religion with specific demographic and ecological variables (Leaf 1972:216ff.). In August I described this and my project to Mr. A. L. Fletcher, then Chief Secretary of the state, who had a reputation for interest in research. In consultation with A. S. Cheema, then Agriculture Director, he in turn suggested two Development Blocks in two separate districts that would most clearly exemplify the statistical trends. One was in an area of large landholders and tenants, the other of independent peasants (a distinction that had not been noted in the census data). Since independent peasants were considered the more "typical" Sikhs in a historical and stereotypic sense (in the literature of the religion and the histories of Panjab), I selected the latter area. It was arranged that the local Block Development Officer (BDO), who lived in Chandigarh, would take me with him on his tours so I could visit a number of villages and find one that met certain further criteria. I wanted, for example, a village of average size, away from major roads, and without recent disturbances, major crimes, or other problems that villagers might want to conceal. Finally, it had to be a village that would be willing to accommodate an anthropologist.

My first introductions in Shahidpur were not conducted by the Block Development Officer but by a teacher at a nearby higher secondary school. The school was situated on a paved road that could be reached by bus (indirectly) from Chandigarh. The BDO and I had stopped there for tea on one of the trips and he had introduced me to the headmaster, the teacher, and several other men. I was, as usual, asked my impressions of the villages we had seen so far. I indicated that while they all seemed interesting, none was exactly right (most, for example, were both relatively new and much smaller than average for the state). The response was that there was a village not far away that seemed to be just what I was looking for. The headmaster's and the teacher's families had originally come from there and had migrated before the Second World War to what became Pakistan. After partition, in 1947, they had returned. There was no longer room in Shahidpur itself for them, so they resettled nearby where land had been vacated by Muslim families. Nevertheless, they still had relatives in the village and we would see them. The village was Shahidpur.

The next day I took a bus back to the school and bicycled about two miles to the village with the teacher, Rajinder Singh. According to custom, we went first to the house of Rajinder Singh's nearest kin and explained the purpose of the visit. Then we went to visit various other "important people" and discussed the project with them. I was heartily welcomed and assured that it would be "an honor" for the village to have such a study done. A room was promised atop a farmer's house, along with all necessary help (an offer that assumed I would demand a great deal in the way of cooking, cleaning, and generally being doted upon). On the strength of all this, but mainly because I liked the appearance and style of the village and the people I had met, I went back to Chandigarh for some needed baggage.

Three days later when I reappeared at the school and asked for help in carrying everything to the village, Rajinder Singh seemed embarrassed. I was told the room was no longer available. A student was now renting it (an obvious fantasy). I would surely not like the food. I would not like the living conditions. It was not really a very interesting village anyway. Wouldn't I be better off staying in a room in the school and studying a village close by on the paved road and busline?

When it was clear I would not be put off, we went to the village. More embarrassment and discussion. A room was finally found in the village gurudwara (Sikh temple). There is a general rule that gurudwaras should provide for travelers, and the villagers evidently felt that they could not say it wasn't available. The room had been used for storage and was empty save for a good half-inch of dust on the floor. I was given what must have been one of the most dilapidated cots in the village: narrow, short, and sagging. A cleaning crew was promised immediately, but throughout the first afternoon and evening no one came. Dinner included a most unwelcome insistent offer of "wine" (a kind of orange-flavored gin). Happily, I accidentally knocked over the tumbler and there wasn't any more to refill it. (I was later told this was considered very clever of me.) I sensed that the offer was insulting, but couldn't say why. The next morning, still no cleaning crew and more strange behavior. That afternoon I insisted on a broom to do the cleaning myself. What appeared was a peculiar sort of whisk broom woven of palm strips (which I used, evidently to everyone's surprise). Several days later, when I still hadn't gotten the idea, I was offered a new cot of the more usual (nicer) type. It was much longer and wider and the webbing was carpetlike (not netlike) in a colorful pattern of home-dyed, soft cotton yarn. Within three days I began to notice itchy bites on my arms and legs but saw no mosquitoes. Upon enquiry I was eventually told, obliquely, that the house from which the cot had come was well known for its bedbugs. Children were bad mannered and noisy. Adults generally maintained a surface politeness, but it was impossible to find out anything unless I could make it appear that I already knew it (a challenging requirement). There was, however, enough support and interest for a start.

Gradually the weather improved and it became easier to think and to work. I learned whom I could and should turn to for help and to represent the mainstream of village ideas and whom I should avoid, and of course the villagers learned about me. The carpenters made me a table and stool for my work. The way I was perceived slowly shifted. I was regarded no longer as a potential threat to the young women of the village (I later found this was the major fear behind the initial withdrawal of the welcome), but as a student of village matters of some real, if unclear, sort. No one went out of their way to prompt me or tell me what to investigate, but most villagers would answer questions and then wait to see what I did with their responses, occasionally examining me on some specific topic or another. There were expressions of annoyance if I stayed on a point too long or seemed to overemphasize or underemphasize it, or if I was too slow to put some facts together in their proper relation. There were expressions of satisfaction if I picked up a point quickly, entered an analysis of some local event competently, or handled a tense situation properly (like being followed by a crowd of rude boys or being given a cot full of bedbugs). A pattern of work developed alternating elicitation in the village with backup work in Chandigarh, as described in the next chapter. But even though a great deal of information was ultimately provided, some suspicion of my purposes remained and I continued to find it necessary to strain every possible inference from what I already had learned in order to obtain new information by already seeming to have it (or almost have it).

Because of the constant suspense and tension, I was quite surprised when, at the end of the last day of my last visit to the village, a few friends I was walking with conducted me past most of the major villagers standing in a line—including all of those who had initially welcomed me and then organized the attempts to discourage my acceptance of that welcome. As each shook my hand, he heartily and forcefully bid me a speedy return and assured me all cooperation would be forthcoming. It was genuinely moving, but what did it mean?

About two weeks later, as my wife (who had joined me after the first year) and I were finishing packing in Chandigarh, two friends surprised us with a final visit: Netar Singh, son of Indar Singh, and Naranjan Singh, son of Bakhtavar Singh. Netar had been in his final year of high school when the study began and even then was a perceptive student of village politics and customs. In cooperation with his father, he had been my main guide and teacher of Panjabi. He had an abiding interest in the ironies of the ways social rules interact with rational calculation and personal character. Naranjan was a couple of years older and a good friend of Netar, though a different sort of person entirely. He was equally reflective and had been equally helpful and supportive in quite different ways. Since Netar enjoyed discussing hidden motives, I asked him what the invitation to return actually meant, reminding him of the first days. He just said, "I do not know. You will have to come back to see."

When I left India in March 1966, there were too many immediate worries for me to give much thought to when, if ever, I might go back. Gradually, however, the other matters resolved themselves and a sense of separation began to grow. This is the other side of Srinivas's coin; to see the village, you eventually become a villager, in a way. You realize that your stay has been part of *your* past, not just *the* past. Your understandings of the village become tangled up in your predictions of your own future, but equally your future becomes entangled with expectations and hopes for the futures of the villagers. The desire to check on your own work cannot be separated from the desire to learn the subsequent fortunes and fates of those whose ideas you made your own. Although I had maintained contact through correspondence, writing to Netar and Naranjan and to other friends in Chandigarh, the questions grew far too numerous for letters.

My understanding of the most general bases of the villagers' concerns and mutual relations was published in 1972 under the title *Information and Behavior in a Sikh Village*. Although copies were sent to the village for comments, none were forthcoming. The monograph concentrated on the ideas that most villagers assume, and say, "everyone knows." The relationships between such ideas and behavior are by no means simple. In the field study, my effort to sort them out focused on an analysis of the weddings that occurred in the village during the summer of 1965. This related the social ideas represented symbolically in the marriage rituals to the "interests" that the central actors in the rituals defined for themselves by applying the same ideas to themselves in nonritual activities. That is, a person who took the ritual role of father, which was built on the general cultural idea of father, might or might not also take the "actual" role of father by applying the same idea to himself in nonritual contexts. Whether the ritual father was the nonritual father or not could be explained in terms of other definitions and of his personal circumstances. For example, in one case an "actual" brother acted the part of the ritual father. The reasons for this were given partly in terms of economic ideas, partly in terms of kinship ideas, and partly in terms of political ideas. Precisely the same ideas were used to explain the relationship between the brother and his sister outside the ritual, and in general the process of assuming the ritual roles represented and duplicated the process of selecting and acting upon social roles normally. On a larger scale, the overall sense of social reality itself was created in the same process and by the same means.

The analyses of marriage ceremonies were left out of the monograph because of considerations of length and neatness. Illustrations of the way the systems of understandings appeared in behavior were provided by smaller-scale common events and by simpler rituals and other stereotyped acts. Published criticism soon made it clear that more had to be said. This raised the question of whether, in order to say it, I should return to the old data or collect new material. It seemed to me that the latter would make a stronger and more interesting case.

The opportunity for a return visit that presented itself in 1978 resulted from a conjunction of the relaxation of personal demands and improvements in the international and Indian political climates. I wrote Netar and Naranjan that I would arrive in the village on the morning of June 12. After many delays in the initial date of departure and some unexpected difficulties in New Delhi and in Chandigarh, the appointed morning finally arrived. At 10 A.M. I sat in a bus in the dusty parking area of the Chandigarh bus depot, writing some notes and rehearsing the old uncertainties. It was late to be starting village business, but I was satisfied to make the appointed day. As a reasonable minimum I hoped to be able to learn how the first monograph had been received, obtain some idea of magnitude and direction of changes in the intervening time, and assess the likelihood of cooperation in a new study for which I wanted to seek funding. At the same time, I would determine local research costs and the level of official cooperation that might be forthcoming outside the village. In 1965 there had been an important pattern of migration out of the village to other rural areas. This had been neglected in the literature of economic and social change (which tended to concentrate on rural-urban migration). If such migration was widespread and had continued throughout the intervening period, it might be an important and unexpected source of the transfer of technology from the more to the less well developed rural regions and thus merit an extended project. At this point, I had still formed no firm opinion on the Green Revolution as such, but was rather inclined to be skeptical.

Along the thirty-mile, two-lane, tree-shaded, blacktop road toward the village, everything looked the same, yet slightly different. This late in the hot season the fields should all have been cleared, plowed, and leveled to preserve moisture. Some crops would have been sown; others would await the rains that were due to begin shortly. Yet many of the fields appeared to have been only recently cleared and were yet unbroken by the first summer plowing. Other fields were full of what appeared to be young rice, a crop almost never grown in this area during my first stay. What had been open sky in 1965 was now laced with electrical wires on concrete poles. Along the road village after village had small sawmills, machine shops, and new retail shops of all kinds. The towns were noticeably larger and busier. The roadside walls, formerly blank and faded, were now often brightly decorated with advertisements for fertilizers, insecticides, and electric fans. On the road itself the confusing assortment of pedestrians, camels, buffaloes, and goats was still there, but it included many more taxis, scooters, motorcycles, and bicycles than I remembered. Everywhere there was activity and animation. Formerly the period between the onset of the hot season and the beginning of the rains had been a time of cautious waiting, a time when the countryside seemed to hold its breath to conserve its strength.

Following former practice, I transferred at Morinda to a local bus and took it as far as the village of Macchli. From there I would walk the mile and a furlong north to Shahidpur. The bus slowed almost to a stop, I swung off and found my footing on the familiar ground. The deep pond beside the road was still there. The shade of the large pipal tree at the far end still looked cool and inviting. But before I could consider sitting there for a few minutes to get my bearings or look for the road to the village, my name was called out. I turned and saw two young men walking bicycles across the road toward me. They introduced themselves. Surindar Singh, son of Indar Singh, had been ten when I saw him last. He was now awaiting the results of his M.A. examination and bore an uncanny resemblance to Netar Singh, his older brother.

Dharmpal, the son of Babu Ram, I did not remember personally. He had been just six when I left and was known to me mainly through the hopes of his father. Babu Ram had drawn a hard lot in life and bore it with widely recognized grace and gentility. In 1965 Dharmpal had been his youngest surviving child and only son.

Dharmpal took my bag. I sat on the carrier of Surindar's cycle and we started off over what in 1965 had been a rough, rutted, dirt track. Now it was a raised asphalt road about four meters wide. My fears had begun to diminish from the moment my name was called. As we pedaled along they evaporated completely. First Surindar carefully and politely, but thoroughly, upbraided me for being so late. He explained that most everybody in the village had waited to welcome me at the gurudwara, but by 8:30 they could wait no longer. They were working people.

As we passed through the fields each new thing was pointed out and its origin explained. We rode past cylindrical haystacks about twelve feet high, of a type I had never seen. They contained wheat straw and the technique for making them had been introduced by Panjab Agricultural University. My impression of the changed agricultural cycle was correct. It was due in part to the fact that the old varieties of wheat, the main winter crop, were no longer grown. They had been entirely replaced by new high-yielding varieties which were harvested later. The formerly ubiquitous Persian wheels that would have been constantly in use at this season, driven by slowing plodding camels or pairs of bullocks, were now almost gone. The few that remained were lying in ruins. Replacing them were mechanized pumping sets in small brick buildings. Small scattered plots of bright green were rice seedlings waiting to be transplanted when the fields were ready—in a few weeks. Overall, it seemed clear that the pattern and rhythm of village agriculture had changed even more than the bus ride had suggested. As a main source of the change, again and again there was praise of the Panjab Agricultural University—founded in 1962 and hardly mentioned during my first stay.

As the information poured out I had to ask, "Do you think it would be all right if I tried to do a study of these changes, to write a second book?" Surindar's reply was "You must." Since such a study was clearly of great importance both practically and theoretically and would not interfere with my plans to test the possibility of the larger study later, it was not necessary to hesitate in replying that I would do it. My decision was reinforced later in the day when I learned that the pattern of migration to surrounding states had stopped and in fact been reversed. (A few days later, back in Chandigarh, I was to learn from Professor Gosal that he had been following the migration story on a large scale, and had it all quite completely in hand.)

We left the paved road and, after dismounting from the cycles, followed a small dirt path that cut off to the left. The last few dozen yards passed through some work areas to the village houses, grouped within what had once been an almost-continuous rampart of adjoining outer walls. Here again was the familiar and new. The approach route was the same: by a line of trees and some tethered buffaloes, we entered the village center through a farmer's courtyard gate, turned to walk past the old Hindu temple and thence into the bricked lanes within the compact mass of houses and barns. The cool, narrow lanes themselves, between high, patinated brick walls, were the same. But there had been new houses near the path on the way in, there were electric lines overhead, and the familiar silence was broken by the chugging of a diesel pump somewhere nearby.

As usual at this time of day in this season, the lanes were deserted. The air smelled of dust and dung, smoke, water, and fresh green plants. We proceeded directly to Surindar's family's house. Word of our arrival spread and the courtyard filled quickly with many old friends and almost as many children born since I had left, all in a holiday mood. After the initial greetings, salutations, and reintroductions, at a moment when conversation seemed to grope for a new direction, I raised the worrisome question of the value of the 1972 book. It turned out that there had been no response because there was not much to say. It was considered quite an ordinary, noncontroversial description of what everyone knew, except for two problems. The first was that I made an "error" about the name of the village; it was not "Shahidpur," although Shahidpur was a perfectly good name. I tried to explain the fears that lay behind the anthropological practice of disguising the places where research was conducted and the concern for anonymity in then-current "human subjects" regulations being proposed by the federal government in the United States. This made no impression. The villagers had nothing to fear and insisted I use the "correct" names both of people and of places.

The second problem was that while I left the village with a draft of the first study in hand in 1966, the monograph had not been published until 1972. Seven years, I was told with much nodding of agreement among the listeners, was much too long a time to account for. The new book should come out much faster. A year, two at the most, should be enough. Again, my explanations about the problems of reviewing and editing made little sense. I promised with misgivings to do the best I could. Privately I was delighted that the descriptions seemed so simple and natural to those described. Even assuming that points had been missed because of language difficulties and that discrepancies would probably emerge in more-detailed checking as I proceeded, it was the reaction I had most hoped for.

Since the monograph was evidently not going to support much discussion, I opened a box of photographic prints from the first study. The time they represented had become history; there was much that the older people present had to explain to the younger. There was, of course, also much that needed no explaining, for part of history never changes. Moda Bhagat had died in 1967, soon after I left. Although he was an "orphan" who lived on a cot in the village gate, was not a native of the village, and was of lower-caste background, he had been a much-beloved and respected figure and prominent leader of village opinion. His picture would now go in the village gurudwara. Most of the other portraits were of people still living and were welcome as reminders of their own pasts. Conversation constantly returned to the way in which events had outstripped everyone's expectations.

As the shadows lengthened, a walking tour of the village lanes was in order. Since school was out for the day, we made a parade, led by adults but with a trail of children of all ages, jumping about and running, alternately taking the part of participants and spectators. Changes continued to emerge, literally at every turn. The village gurudwara had been substantially enlarged, electrified, and supplied with ceiling fans and fluorescent lights. New housing was everywhere. The houses too were electrified, and most, I was told, had radios, fans, and lights. There were new bricked lanes and houses where open fields had been, two new doctors' offices, and a new rest house for visitors on the "Harijan side" of the village where the lower-caste groups had traditionally lived (the new housing was not segregated on the old pattern). A new middle school had been built on what had been in 1965 one of two separate cremation grounds (there had been one each for the "Jat side" residents and for those of the "Harijan side"). The remaining cremation ground was now used by everybody. Indar Singh (Surindar's father) had been a prime mover in the effort to get the school and as we walked by, I asked him about this change in the funeral traditions. The answer was direct. "Education should be the first priority," and in any case "in death everyone is equal." The school had opened just three months earlier, with a staff of five and eighty-one students, in grades six through eight. The students came from Shahidpur and about four nearby villages.

Still later in the afternoon, Netar Singh arrived from his work place 200 kilometers away in a taxi (a mode of transport formerly reserved for the bride and groom in wealthier weddings). At first sight I did not recognize him. He was about thirty pounds heavier and several inches taller. He was now a husband and father, and as a grain inspector working for a new marketing cooperative he had access to information and ideas that were obviously of considerable local interest. It did not take me long to see that he had blossomed into a kind of storyteller and political commentator and that his perceptions and judgments were respected.

We continued the tour together and just before sundown wandered back toward the well and work building of Indar Singh. Since well sites are regarded as private, it was possible to dismiss the train of children. We washed in the cold, fresh water pumped from the well into a large concrete watering trough and sat quietly on cots under the trees continuing the conversations of the day and of the years before.

The new crops in the village reflected improved market and transport facilities. It was explained that whereas formerly the nearest market had been in Kurali, seven miles away by dirt track, now there was a market in Uncha Pind, just over a mile from the village by asphalt road. In fact, I was told, I should no longer use Macchli as a bus stop, since Uncha Pind was just about as far from Shahidpur and from there buses to Chandigarh were direct. At the same time there were always people coming from Uncha Pind to Shahidpur who would give me a lift, or I could borrow a cycle from a tea shop (which Netar would arrange) and pedal myself.

As darkness fell, the group broke up still further. Surindar, Netar Singh, and Indar Singh took me and a few other friends to their house for dinner, where the surprises would continue.

In Indar Singh's house, Arjun, Netar's elder brother (who had been working and was not with us at the well), asked me if I ate meat. It was a surprising question. Previously meat had been eaten only at weddings (and I had eaten it along with the others). Families planning a wedding would usually raise a goat or two against the occasion. Accordingly I said no (not wanting the goat killed just on my account), unless they would otherwise be eating it. Arjun explained that the question was not whether I demanded a goat (regardless of trouble or expense). Now, he said, a goat was killed somewhere in the village on almost any night; it was only necessary to go and buy some of the meat. The family normally ate meat six or seven times a month and they felt like having it that night. (It was clear that the reason for this "feeling" was a celebration of our reunion, but this was not said explicitly.) Would I join them? Since it would not be an imposition, and since it was the truth, I naturally said I ate it. Arjun disappeared and returned shortly with a small paper packet.

The formal pattern in Panjab villages is to feed a guest alone and in general to feed the men of a family one at a time. The attention shows respect and also allows each person to be supplied with the *roTī*—the flat soft breads that are the staple of the meal—as they come hot from the fire. On this occasion we were less formal. I was seated under the covered portion of the interior courtyard, in a backed chair, facing the open area. A folding steel TV table was set up before me. Close on my right, Netar Singh sat cross legged on one end of a cot. His back was to the wall as mine was; he too faced the open courtyard, across which his mother worked at her small hearth behind a low wall. On the other end of Netar's cot, facing him and at an angle to me, was Naranjan Singh, back from his day's work as a billing clerk for a company dealing in tube-well equipment in Morinda. Just beyond the foot of the cot was a large electric oscillating fan on a small table.

Naranjan, thin as a rail and unchanged since the first day I saw him in 1964, had joined us at the well. He had walked with us to the house, steadfastly refusing repeated invitations to join us for dinner. There was "no need." He then equally steadfastly did his share of damage to the large stack of *roTī*. Most of the other family members sat about on low stools, door sills, and other cots and joined in the conversation. The meat was excellent. Obviously Netar's mother was familiar with its preparation. It was now quite dark, but on the wall above us, under the roof, a lone yellow bulb with a small porcelain reflector did cheerful service. Netar's mother watched and listened from her fireplace with an expression of happiness and pride—in all of us, I am sure. Much was new and most was better. More changes in all of our lives would follow and we all knew it. But for the present it was good to be back together again, all alive and well, to appreciate our histories thus far. Perhaps too it was good to see how friendship had lasted over the years. For whatever reason, it was hard to imagine a happier house that hot summer night.

After dinner it was almost belaboring the obvious to raise again the question of the desirability of a new study of the village, but it was important to be sure that the agreement was made in full understanding of the work that would be involved. Everyone again assured me that the villagers would certainly like it and would cooperate fully. Naranjan and Netar would personally help with the census and the survey of social relations. We reviewed the rest of what would be needed and they felt sure that it could all be done. Conversation again turned to other matters.

It was after eleven when Naranjan left and I retired with Netar and his younger brothers to the roof. The women, Arjun, Santoch Singh (Indar's brother), and the children stayed below to sleep in the courtyard. Indar would sleep by the well. This was the usual arrangement for this time of year, and the roofs catch a pleasant breeze, no matter how hot the day. Fourteen years before I had lain awake almost all of the first night listening to strange wails and piercing cries, awaiting attacks by expected swarms of mosquitoes, and wondering what new tensions the next day would bring. It had taken many weeks to learn to sleep peacefully in the village. Now I knew the wails and cries to be the calls of peacocks and lapwings. Mosquitoes, at this season, were killed by the heat, and the feeling of being exposed, isolated, and lonely was replaced by the awareness of having returned among friends. It had been an eventful day and would be a pleasant night. For a while, Netar and Surindar carried on a conversation about working conditions with a young woman from an adjoining house, visiting from her job as a medical technician in a city nearby. Above, the full moon shown through rows of silver-fringed clouds. Netar said it would rain when they closed in completely, perhaps in a day or two. I looked forward especially to the cool, quiet time just before sunrise.

It seemed that my eyes had barely shut when I was startled awake by one of the most God-awful rackets I can recall hearing—certainly at 4:30 in the morning. It persisted and began to resolve itself as I struggled to get my eyes open. It was music; it was the Sikh morning service; it was the *Āsa dī Wār*—Song of Hope—being broadcast from the village gurudwara over a public address system of admirable power, and it continued for two and a half hours until full dawn. When the sun rose, I could see that most of those on the roofs continued to sleep peacefully. Netar explained, when asked, that there was nothing unusual about this particular morning, or Shahidpur. Before the first light of every dawn, the entire Panjab was, one might say, ablaze with song. In retrospect, no one thing seems to sum up what has happened in Panjab as well as this.

The service was composed of live singing, taped records, and radio broadcasts. In a broad way this nicely signified the persistence of many traditional religious ideals and values (such as the injunction to recite the prayer each morning before rising), and the thorough and comfortable integration of new technologies into the patterns of village life. But this was only the beginning.

Within the framework of the village ecology, the Song of Hope reflected the broad replacement of biological energy by electromechanical energy. The broadcast depended on electricity both for amplification and for the light that the *granthī* (attendant) needed to work, and of course the purchase of the electricity and the electrical equipment depended on the increased wealth generated in part by these same new power sources. In 1965 virtually all power in the village had come from human or animal effort, fueled by food grown locally. Heat had also come from locally grown fuel and light had been provided by small kerosene lamps, small cans with rope wicks sticking out of the tops giving about half the illumination of a wax candle. Exceptions had been rare. A few villagers had owned dry-cell flashlights. My own kerosene pressure lamp had occasionally been borrowed for nighttime readings of the Sikh scriptures. Once a year at an annual fair a public address system had been used, but the power had been obtained from wet-cell storage batteries that had to be charged in a nearby town and brought to the village with the rest of the equipment.

The amplification of the morning service in 1978 was explicitly described as another labor-saving device. It was no longer necessary for every villager to lay awake and say the prayer for himself each morning; one could do it for all. The remark was whimsical and ironic, but the irony rested on fact: the power used in amplifying the service had indeed enabled a few to do the work of many in many contexts and made life considerably better for all. It is hard to say just how much improvement was involved in the fact that the village lanes were now lighted and students no longer had to struggle to read by dim kerosene wicks, late into the night. It is easier to gauge the importance of drastic shifts in the animal population due to the reduced need for draft power that will be described. But the main point is that almost nothing in the village system of production remained untouched, and electrification was only one thread among many in these ecological changes.

The morning service also reflects numerous interacting trends from the point of view of the village system of management, what I call the division of labor. The broadest is exemplified by the life history of the singer himself—the *granthī* who assembled the service and tended the gurudwara. This trend can be called professionalization and can be described generally as the increasingly intensive concentration of each individual on fewer and fewer forms of work, which they do more and more competitively and with more and more capital. Several families that were farming in 1965 had withdrawn from farming by 1978, divested themselves of land, and concentrated on other professions—such as teaching or other salaried work. Those who stayed in farming enlarged their holdings, greatly increased capitalization, but specialized in management activities, hiring labor for the occasions when it was needed. At the same time, many nonlandowning families specialized in complementary ways. Artisans specialized and concentrated their activities. Others, who formerly had worked part time at different tasks as opportunity arose, became full-time, salaried agricultural workers for specific families, working on yearlong contracts. The pool of those only partially employed in the village, those who would provide casual labor in periods of high labor need, like the harvest seasons, shrank. As a consequence, by 1978 a market had developed for migrant labor which I had never heard mentioned in Panjab in 1965. People came from nearby states and moved from village to village in peak seasons as organized crews, supplying the needs that could no longer be met out of the local population. The *granthī* had come from a poorer area to the south as such a migrant. He worked for ten years in and around Shahidpur and gradually acquired an interest in Sikhism. The villagers asked him to begin serving as *granthī* in 1977 and it is fair to say that, as compared to his predecessor at the time of my first stay in the village, he conducted himself in a more professional way in a very specific and clear sense—working harder to spend more time doing things *granthīs* are supposed to do (according to published standards, in some cases) and avoiding things considered unrelated to his role or downright inconsistent with it (such as growing vegetables and raising goats in the gurudwara yard).

Economically, the morning service obviously provides evidence of the greater amount of wealth in the region, as well as the wider range of things to buy. Cassette recorders, public address systems, lights, fans, and the many improvements to the gurudwara had been either entirely unavailable in 1965, or prohibitively expensive from the village viewpoint. But there is a point of greater significance here than just the increased quantity of goods available and the new wealth to obtain them. Why should the wealth be associated, in the villagers' views, with the gurudwara? Panjabis have not become so wealthy that they can give to any public charity whatever. The support of the gurudwara actually reflects a specific judgment by those who give it, a judgment that the religion itself has been responsible for their fortune, and this judgment is correct. At the individual level, the ideas of the Sikh religion and the common identity of the villagers as Sikhs have been used to form many of the bonds of confidence upon which the new and complexly interacting ventures have depended. There is a widely recognized logical "fit" between the ideas of the religion and the innovative, risk-taking attitudes that have contributed to the changes—the respect for work in its own right, the willingness to make a personal judgment under new circumstances, the ability to trust the sense of moral responsibility of others, and the courage to act on the combination of judgment and trust. The religion emphasizes the importance of the individual, the importance of truth, the importance of "self-sacrifice" in the sense of putting one's obligations to the community above the obligations to one's own person and family. Some would argue that Sikhism is especially well suited to this; that the values are more clearly expressed here than in other Indian religions. I do not think this view would be supported by thoughtful Sikh religious scholars. The basic values of Sikhism are shared by many Hindu schools and go back to the Vedant of the Upanishads. There is no real reason to argue that Sikhism is uniquely suited to such applications; it is enough that it is the predominant religion here and it has worked here. This is what Panjabis, especially Sikhs, widely believe, and it is one of the things signified by the support for the gurudwara.

The abstract importance of the religious ideas at the interpersonal level reflects a much more concrete connection to social and economic development at an institutional level. Revenues from major Sikh shrines in India are gathered together and managed by an elected committee. The committee, and the system of bodies that produces it, has been instrumental in formulating many of the governmental policies that have led to the changes, beginning with the partition of erstwhile (post-1947) Panjab into the two present states of Panjab and Haryana. The support for the local gurudwara was seen by villagers as an aspect of their support for this larger system of institutions, as well as for the abstract ideas of the religion as such.

In a strictly religious frame of reference, the morning service reflects the continuing viability of certain very specific ethical and religious concepts enunciated in the prayer itself and reinforced in the setting from which it emanates. Two of the most important ethical ideas of the Sikh religion, closely related to the individualism already noted and repeatedly stated in the Song of Hope, are the fundamental equality of all religions and the fundamental equality of all people. Consistent with these ideas, Sikh religious teachers since Nanak (whose poetry makes up a major part of the *Āsa dī Wār*) have rejected the authority of Brahmins (or any other purported group of priests) in favor of each person's own reasoning, have rejected any claims that the teachings of one religion are more correct than those of any other, have stressed the importance of good works and leading a good life over the performance of any particular ritual observances as such, and have denied any religious authority or recognition to caste on the ground that God, as the basis of understanding, is equally present in all people and equally available to them through dutiful and thoughtful living.

It is often said, by way of disparagement, that the Sikh religious organization in Panjab is Jat dominated. That is, it is dominated by members of the Jat caste, which is the predominant landholding caste in the region. The disparagement lies in the suggestion that such Jats practice caste discrimination. This suggests in turn that the religious rejection of such discrimination is not viable and implies that the ethical system as a whole lacks real support. It is certainly true that a large portion of the leading figures in the religious organization is Jat, as indeed are those in the state government. This is hardly surprising, since Jats are by far the largest single caste group in the region. In the rural areas they make up perhaps a third of the total population and up to half the population in many villages. Yet all religious leaders, Jat and non-Jat, support the ethical values, and what is true for the state is equally true for the village. The informal committee that runs the Shahidpur gurudwara is predominantly Jat, although the village panchayat (governing council), which has had a role in raising some of the money for its improvements, had a non-Jat majority in 1978. The *granthī*, in addition to being of non-Sikh origins, is in fact of Harijan (lower-caste) background. Thus the management of the gurudwara, as reflected in the morning service, represents not only the continuing viability of these fundamental ethical values, but also their connection to the general openness of Panjab society to both geographical and social movement.

Villagers generally agreed that education and educational institutions, the Sikh religious ideas and religious institutions, and the new technologies were three of the main pillars of the changes that took place between 1965 and 1978, although they understood these as involving much more than has been said or suggested so far. Just how much more will be explained in the chapters that follow, but still by way of introduction at least one more facet of the ritual, one more major frame of reference in which it is to be interpreted, should be mentioned here. This is politics.

Before my first stay in Panjab, one of the areas in the literature that I found most confusing concerned what was called "factions." Most often they were treated as "cleavages" in village structure (cf. Beteille 1965) rather than as organizations in their own right. It was difficult to imagine such an entirely negative entity. How was one to join a cleavage? But more than that, why did they recur again and again, even though what they "cleaved" (caste, kinship systems, economic classes) often varied greatly. To make the picture even more confusing, it was often recognized that village factions had linkages to regional and national political parties; indeed, such parties were often represented in villages by factions. One seldom, if ever, found national political parties represented in villages as such—by local officials, offices, and the like, as in the West. How could a negative organization, a nonorganization, aggregate to make up a most definitely positive organization on a larger scale. It was as though a plate could be made out of a large number of cracks. My guess was that something was being misstated.

After my first arrival in India but before beginning work in Shahidpur, I looked for occasions to ask about politics, and it became clear that it was a much more sensitive topic than in the United States. As for factions themselves, two points became obvious. First, they were discussed not as a *fact* of village organization, but as a *problem* with it. This did not necessarily mean that they were a problem in fact, but it did raise a warning: tread lightly. The second point was that "faction" was obviously *not* the indigenous term for whatever they were.

After I began work in Shahidpur, I learned that the word I wanted was apparently "parties," the English word, evidently, but thoroughly naturalized. Later, Netar made a general observation in passing that has since proved invariably correct, when understood: "Always in Sikhs there are two parties." This does not mean that parties are religious bodies. They are not. It means that among those who are Sikhs, there will always *also* be division into parties. Parties are defined as groups, generally two, of opposed individuals struggling for dominance, and such groups literally have no place in religion, strictly speaking. But wherever there are Sikhs there will be disagreements over matters that arise in the course of religious life, matters that cannot be discussed and decided upon within the framework of the ideas of the religion proper. The religion is concerned with describing the nature of God and the relation of the disciple to God, and with describing certain ethical ideas that follow from it. The understanding is that such ideas are the basis of unity and harmony among Sikhs, not a basis of divisiveness, so that even if it were logically possible to use them in organizing conflict, it would be improper to do so. Accordingly there is no way within a strictly religious context to dispute matters like finances, management issues, and other such everyday problems. This is where the party ideas can come into play—and do. There were two parties in this sense in 1965, and two in 1978. At the state level the parties divide the gurudwara administrative committee. In the village they run through most public organizations. It was precisely because parties were *not* religious groups that they were instrumental in linking religious values to many technically nonreligious topics: secular education, local politics, agricultural policies, and the rest.

Within the village, party ideas were used in organizing support for the improvements to the gurudwara, in building the new middle school, and even in hiring and supporting the new *granthī*—among many other matters. In this sense, the way the Song of Hope reflected the effective organization of conflict through party alignments within the village represented on a small scale the complementarity of party and religion that was so important in advancing secular versions of Sikh religious ideals on the state and national levels. The old ideas of party and religion were used together to create the new circumstances and they thereby imparted to these circumstances the vitally important qualities of controllability and continuity with the past.

The Song of Hope, each morning before dawn, does not by any means represent everything that has happened, but it does summarize enough to indicate both the continued viability of ancient ethical and organizational concepts and the way they have provided a moral framework for the technological and economic modernization that astonishes even those who participated in it on a day-to-day basis. The facets of the ritual are separate. None represents a meaning more "real" than any other. Every event that helped bring about the changes in the village is in some way parallel to the facets of the ritual, and on a larger scale the "Green Revolution" itself is an event of this same complex sort.

Before the Green Revolution, there were many in the West, typified by Gunnar Myrdal (1968), who said such a pattern of change was impossible. According to them, the problem was not one of resources or technology but of attitudes and values. To have self-sustaining progress, these traditional attitudes and values would have to be stripped away and replaced by "modern," "rational," and, in effect, Western values, ideas, attitudes, and organizations. This assumes that what is traditional is not rational. It was a dismal prescription for India, for there was in practical terms no way the thinking of a population of over 400 million could be so altered. Myrdal's prescription rested on a failure to grasp the nature of rationality, its relation to tradition, and just how basic such traditional ideas really were. To leave them is not to become "modern." It is to rip the present from the past, to cease to be whatever one has struggled up to that moment to become: a good student, a *granthī*, a father, a resourceful man, a pious man, a gentle man; or a self-sacrificing mother, a respectful and good daughter, a patient and resourceful wife. It is not to move from traditional values to rational values, but rather to move from reason to chaos. Yet for the world at large, the prescription was, if anything, more dismal still. In their daily commerce with each other the people of Panjab, and South Asia more generally, preserve and pass on one of the world's great cultural legacies. Myrdal was saying that the world could choose starvation with preservation of this tradition, or economic development with its loss. The Song of Hope, broadcast each morning across Panjab from hundreds of villages, presents a more cheerful prospect. It proclaims that the world can preserve these traditions and have progress as well; tradition and rationality are not opposed bases of action but mutually interdependent, and rationality does not lie in the substance of certain values but in a way of using ideas of many sorts, be they traditional or new.

2

CHAPTER

Social Order and Social Change

The Green Revolution can be construed in a narrow or a broad sense. In the narrow sense it consists primarily in the adoption of the new high-yielding varieties of wheat and rice and associated technologies. In the broad sense it includes not only this but all other economic changes as well as the social and cultural changes that either contributed to the technological and ecological changes or were derived from them. The broad sense is what concerns me here.

Many anthropologists and sociologists argue that no scientific theory can describe an entire community at a point in time and also trace changes over time. Any such statement of the limitations of social science can become a self-fulfilling prophecy, but this does not make it correct. There is little point in engaging in abstract argument against such a view or in criticizing the unnecessary and unwarranted theoretical and philosophical assumptions behind it. The only really decisive argument is to get on with the job. This requires a pause only long enough to explain the three main theoretical ideas that underlie the field work this study rests on and the organization of the presentation. These are the ideas of the community, information systems, and communication. Decision making, which is of special importance in understanding social and cultural change, is a subcategory of communication and follows the logic of information systems.

THE COMMUNITY

The simplest way to think of the community is as the total assemblage of objects, living and nonliving, natural and artificial, in the area under study. Objects exist in time, and change in time, so time is an inherent aspect of the community.

The objects in a community are inherently ambiguous, precisely in the same way described for the objects associated with the morning singing of the Song of Hope. Every object has multiple possible social and cultural identifications, and no object has any one social identification inherently or essentially. An uncle is always also a brother and a man, and may be a carpenter or an architect or a fool. A building may also be a gurudwara, or it may also be a public legacy, or it may be on fire. Action toward it changes according to the identification, but every identification must first be asserted and accepted. The same is true of any physical process or any long-term pattern of objective change over time.

INFORMATION SYSTEMS

Information systems are sets of interrelated ideas, established in consensus. Several such systems—never just one—are always present in any community. They are the basis of communication and by the same token they are the sources of the ideas used to interpret the objects in a community. The frameworks that defined the several aspects of the Song of Hope are information systems.

Information systems should not be thought of as purely mental objects, nor should it be believed that they are "subjective" or difficult to describe. They are taught in and through physical activity and learned in and through physical activity. The activity ranges from speech and pointing to long-drawn-out demonstrations, but all types are quite observable and indeed are often far more easy for people to observe and agree upon than "objective" physical phenomena as such.

Information systems are often conveyed by members of communities in the form of descriptions of various realities: of how one grows wheat, of how one calculates return from farming operations, of a family, of a Sikh, or of a party. This is why anthropologists have often construed them as *mere* native opinion about the things they appear to describe, as "native conscious models" that may agree or disagree with native realities. There are, of course, real matters of opinion in any community—the opinions of this person or that. There are genuinely "mere" models, in the sense of speculative constructions about things the speculators themselves know are generally understood to be different from what they say. But there are also ideas that are not merely about things but fundamentally constitutive of them, and these are what make up the basic information systems as distinct from messages constructed with them.

Many of the most important "things" in any community are nothing but ideas in use. To get at the main form of these ideas, one has to take into account both the things people say they believe and those they say they do not believe and find the assumptions underlying both which they cannot conceive of rejecting. One has to sort out the formulations that vary from person to person to arrive at the ideas that are accepted as basic assumptions in large interacting groups. These are the information systems in the present sense. The affirmed and rejected beliefs and the personal opinions are personal elaborations upon them; the formation of such beliefs and opinions is just one of the many uses to which the more basic assumptions can be put.

Every distinct information system has a distinct logical structure which lies in the number of ideas that make it up and the rigor of the logical interconnections among them. Information systems are not exotic objects. They are what we learn when we learn any trade, language, system of calculation, religion, or indeed anything at all except a single word in isolation (if there can be such a thing).

The structures of information systems have to be elicited as an empirical matter, and we are all familiar with the process; we are trained in it in an informal way as a result of growing up in a human community. Gradually we learn the key ideas, and how to relate others to them, as we build up our knowledge of any system in any capacity: whether as children in families, students in schools, apprentices in professions, or anthropologists in the field. We learn that in some systems definitions are all important and cannot be changed. In other systems, observation and flexibility are more important than definitions. We would not change our ideas of God or father because of the behavior of the world or of a specific father; we would not retain our idea of flammability if we found something burning after we had been told it would not.

The logical structure of information systems translates directly into a structure of probabilities, as described in Shannon and Weaver's *The Mathematical Theory of Information* (1964), and this has direct implications for their use in creating both social order and social change, as I have previously described (Leaf 1972: Chapter 1).

Basically, systems of information with loose formal organization, many elements not tightly related to one another, correspond to what Shannon and Weaver describe as information systems or "message sources" with high information potential, while information systems with few elements in more rigid association correspond to message sources with low information potential.

Information systems with loose organization, according to Shannon and Weaver's reasoning, are sources where any given element has a very low probability of being chosen to be used in a message and where, given any one choice, there is still a low probability that any other element, idea, will be chosen next. These are systems where one has high freedom of choice in constructing messages. By contrast, in systems with few elements any given element has a high probability of being chosen in the first instance and, given any choice, the probability of any of the remaining elements being chosen next is even higher. These are systems that allow low freedom of choice.

Shannon and Weaver argue that the directive force of a message in behavior is directly related to its probability (which is the product of the probability of all the independent choices it reflects). A message that reflects one choice out of thousands can be used to key a response that will be similarly distinguishable as one out of thousands. Such a message and response will inevitably appear very particular and precise, even though the message may be very short ("grow two acres of wheat this winter") and the action sequence that corresponds to it and shows that it was understood may be very long and involved. This is precisely what Shannon and Weaver mean by high information potential. On the other hand, if a message only represents one possibility out of three or four, the response it can call forth can similarly only represent one of three or four; hence such courses of behavior can have only limited specificity.

Each person can only learn a few information systems, and in any community only a few are widely shared and form the basis of the most widely accepted interpretation of the objects of common collective importance. In Shahidpur there are six such systems, as already indicated, and they represent a broad range of differences in formal organization and information potential, as will be described.

There are many other information systems in use in the village, but they are taken as defining specialized knowledge that is not available to everyone and/or that everyone is not called upon to use. There is, for example, the system of rules for record keeping used by the village clerk; the medical knowledge of the doctors; the professional knowledge of the shopkeepers; and the trade skills and knowledge of the potters, carpenters, miller, and the like.

An obvious question is why there are, in a community, multiple information systems of different levels of information potential. Why doesn't the more precise and powerful system supplant all the others? This may suggest a still simpler question: if low information systems do not have directive force on specific courses of behavior, why do they exist at all? This is a crucial question and the answer is a major thesis of the present work. As already suggested, the systems with high information potential (ecology, management, economy) define specific actions. But in the nature of the case, precisely because the ideas that define them can be used with so much freedom, these actions do not form logical patterns or groups in and of themselves. As the ideas that define them may be freely associated with one another, so may the acts as defined by the ideas. Plowing wheat is plowing wheat, and one who is plowing is one who is plowing—no more. One who plows may or may not harvest, may or may not make pots, may or may not own the field. Such a label (one who plows) conveys no rights, has no long-term moral significance.

The systems with low information potential, by contrast, define general roles and relations that can be applied in many different physical settings precisely because they lack direct technical significance. A father is stipulated as always a father, a mother always a mother, and it doesn't matter if the father is walking or standing, plowing a field, selling a quintal of wheat, or making a pot. Hence the idea of a father can be used to tie together and give a common significance to many such otherwise separate, technical acts. More generally, when the two kinds of ideas are used together to impute order to objects, including people and what they work with, the low information systems provide labels that give long-term significance to actions precisely because they can apply to the same individuals in a large number of discrete technical activities, and thereby permit those who apply and accept the labels defined in those low information systems to see the activities as done by "the same" person in "the same" set of relations to "the same" others. Conversely, the high information systems allow people to pass on specific instructions for engaging in activities which will allow each person to obtain a livelihood.

All this weighs directly against the arguments of those, like Myrdal, who think that one cannot have technical or economic change without change in basic social or cultural relations or religious values. It means that one *cannot have* directly correlated change in systems of high information on the one hand and systems of low information on the other. One *cannot have* change in the religious system that is directly correlated with change in technological or economic behavior. The systems are too unlike each other, too different in their logical functions, for the idea of such correlation even to make sense. Because the logical structure of each system is so different, each system must change in its own way, at its own rate, and this is what we see in Shahidpur. Moreover, we can expect change to be both less disruptive and more rapid in the systems of high information potential, precisely because the elements can be easily separated and are more tightly related to specific actions than to each other (see Leaf 1972: Chapter 9 for a fuller explanation).

The systems of low information potential, such as the religious ideas, cannot be changed in pieces. So change is therefore more likely to be wrenching if it occurs—and wrenching change is more likely to be resisted. Each individual at each point in time is faced with changing the entire familiar framework, or not changing it at all. At the same time, precisely because such systems lack definitive behavioral significance, there is little reason for them to change, as indeed they did not in the present case. They do not change at all in fundamentals over very long periods of time—not just decades but even centuries. What does change is consensus on specific applications of some key ideas, but even these subtle changes come about only as a result of large-scale consensus-building activities, reflecting much effort, as will be described.

A second point is even more important, and is one that emerged most strongly from the events in Panjab between 1964 and 1978. It is that stability in the low information systems can and probably must be a *basis* of change in the high information systems.

Some anthropologists may be inclined to see information systems as "taxonomies," mere sets of labels, and therefore to see their use as a matter of selecting a label out of a system to "apply" to some object. But it runs much deeper than this. The ideas do not merely "apply" to objects like a coat of paint or a name tag; but are imputed to them as underlying realities that the objects themselves, as seen from moment to moment, merely represent or "document"—in a conceptual procedure that Harold Garfinkel has described as the "documentary method" of interpretation (1967). That is, subjectively the idea is not seen as standing for the object, but rather the object is seen as representing the "thing" (the idea as imputed). This is not a set of labels; it is literally a set of relations of mutual relevance among culturally established realities.

COMMUNICATION PROCESSES

Whenever anyone imputes an idea from an information system to any object in a community, whether in a ritual, in conversation, or in private conscious thought, communication takes place. Technically and properly Shannon and Weaver's formulation requires that a message be created by a sender and interpreted by a receiver, but the theory recognizes that the same physical person may serve both functions. That is, conscious thought is also a form of communication. So too, more obviously, is decision making—both the private calculation and the public action necessary to implement whatever choice is made. Communication includes, literally, the formation of any meaningful behavior whatever.

METHODS OF OBSERVATION

Information systems are not difficult to find, although the discipline to record exactly how they come to our attention and to trace them out systematically may at times be hard to sustain. Often they are literally forced upon us. We have had this happen to us all the time as natives of our own societies, and anthropologists experience it regularly when studying other communities, although they usually reflect on the process under different labels and through the distorting glass of theoretical commitments that bid them construe what they are doing as observing "behavior." In reality, one cannot see behavior without using one's ears: one looks and also asks what it is that one is seeing. One is invited to attend something and told what it will be. Thus, however one may twist to construe it otherwise, one obtains information on ideas and their imputation to objects in the community. This basic elicitation procedure, usually called participant observation, was supplemented in my first study by two other procedures: direct elicitation of the information systems as such and surveys to determine the distribution of major imputations.

For the more highly organized systems, direct elicitation involves identifying a few key ideas in advance (by seeing what is imputed in behaviors), finding the setting in which the ideas are used and the type of group that uses them, finding a way to be in such a setting, and then simply being in one and finding an opportunity to ask for an explanation of the key ideas. In the case of the Sikh religion, for example, before I ever began work in Shahidpur I was taken to a recognized scholar, Sardar Balwant Singh, by Professor Indar Pal Singh, my initial host in New Delhi. Balwant Singh in turn began by repeating the main themes I had already noticed in talking to others and reading the literature. There were just two key ideas: the Sikh and the Guru. I then asked about them, and he explained their mutual relations in one very concise lecture of about an hour. When he finished this, I asked about the relationship between these major ideas and a few themes he had not mentioned up to that point, mainly those revolving around Sikh militancy, and he showed how they were connected in another hour. Everything I have encountered since that has had wide support in consensus, in the village and elsewhere, has been consistent with this initial exposition. Similarly, it took just one long session of several hours with a large family group to elicit the main kinship terminology (although several other sessions were used for checking).

The systems of high information potential, by contrast, cannot be elicited all at once from one or a few authoritative sources. Instead, one has to look at each specific set of major activities (working out a procedure for visiting a sample of work areas in space and time to be sure to touch on all of them). Then one asks for general explanations of the procedure in question—but not of the system in general, unless one has very good reason to believe the question will not lead the respondent into areas that are not actually established in firm consensus. If the general explanations overlap and form a consistent whole, well and good—one can put them together and check back with the informants to be sure the resulting construction is not arbitrary or silly. But if the explanations do not overlap, one must always be prepared to consider the possibility that there are several distinct systems of ideas involved.

Basically, in either case, one looks for ideas that are *invariably associated* with one another and are established in *substantially uniform consensus*. These are the ideas that form the baselines of understanding. They are not necessarily what everyone will say they believe most firmly; rather, as already suggested, they are what substantially everyone says and assumes everyone *else* believes and what they describe their personal beliefs in relation to. They define, in a literal sense, the common sense of the people of the community.

Once one has these ideas in an orderly form, it is necessary to take the next step and ask which ones each person uses to define his or her own most basic concerns and relations, especially those acknowledged by others. Other more casual imputations are not to be ignored, but in the long run it is the success or failure of people in managing the ideas used to define their most immediate interests that will determine whether the overall pattern of community life stays the same or changes. If all those with camels decide, for their own reasons, to change the way they treat them, or to eliminate them (as happened in Shahidpur), the ideas of a camel current in the community as a whole will shift. If those people considered most representative of the ideas of the Sikh decide to emphasize one idea and deemphasize another, and if they obtain broad public support (as happened in Panjab as a whole), the ideas of the religion are changed for all.

In the 1965 study of Shahidpur, such data were obtained in three major social surveys and from the official land revenue records. The first and most important survey was my census, administered to every house. The first section asked for the house number, number of house units (freestanding buildings or plots), location of units, and number of rooms in each. It asked how long the unit had been occupied by the family, and it asked if there were any units with purchased rather than inherited rights of occupancy. A section on house ownership asked the owner's name, caste, clan, and whether he was a landholder. It asked if the owner was head of the occupying household and, if not, what his relation to the head of the household was. Finally, it asked if the house was ancestral (inherited) or self-acquired property of the owner. With respect to the occupying family, it asked the name, caste, and clan of the head, and whether he was a landowner. For all others in the household, it asked name, age, relation to head, where they ate, where they slept, and whether those arrangements were permanent or temporary. It asked if there were related households in the village, and for the name and relation of the head and the house description and location of each. With respect to the contents of the house, it asked for a list of any "special equipment" (*khās sand*, which would mean the tool that would be associated either with one's caste occupation or more generally with any major occupation—such as the loom of a weaver or the shop of the shopkeeper). It ended with a checklist of all types of cattle kept locally, and asked each household to describe the number and location of each type which they actually owned.

In going house to house to take the census, I usually stood at the house entry or sat just within and spoke to the woman of the house rather than the man (who was out working) while a crowd of children gathered about as onlookers and translators. The translation was needed because women were generally not literate and spoke a more rural type of Panjabi than the men. In this they did not break sentences into words. Since I had received my Panjabi instruction mainly from men and could not break my English-speaker's habit of pausing between words, it was difficult for the women to understand me. Their speech, run-on, different in vocabulary, and most often spoken looking away and through a shawl, was absolutely impossible for me to follow. So I would ask my question in my Panjabi. The children, who had been to school and knew the convention of dividing sentences into words, would restate it. The woman would answer in her style, and the children would restate it again in my style. The children also stood guarantee as a kind of chaperon.

A supplementary form going into more detail on additional social relationships with other households was made up to be administered to any family that would spare the time. The first section asked for the name, duties, and payments to all those in other castes with whom one had *lāgi* relationships. The term *lāg* is associated with intercaste payments for household service and corresponds to what ethnologists have described as traditional payments among hereditary patrons and clients in the "jajmani system" that is widely described in the Western ethnological literature (cf. Wiser 1936; Beidleman 1959). The second section asked what work the respondent himself did on contract (the same idea—by the time I designed the form, *lāg* had already been described as a kind of contract payment). The answers to these questions quickly demonstrated the marks of imputations without sanctioned acceptance. There was very little agreement on what the traditional work was for all but two or three castes, and the reported payments not only varied widely from one respondent to another but included such implausible formulas (given the usual way bargaining was conducted) as "and whatever else he asks for." Also, while people were often willing to say what they gave, no one admitted receiving any such payments. Further, if one added up the payments supposedly given and compared them to the evident incomes of the supposed recipients, there was no correspondence.

The third section asked who was in the respondent's *melī* (the group expected to come to weddings and eat with one's own family, and that should support one's family if a conflict arises).

The fourth section asked for the name of the head of the family that each responding family had actually given wedding support to or received wedding support from in the past three years.

The fifth asked who the respondent's close friends were. I had been told by Netar, Indar Singh, and others in a pretest that the "friend" question (suggesting parties) would not be one people wanted to answer, and this proved to be true. The usual response was discomfort, a question about why I wanted to know, or a request for clarification. It was clear that the question was inappropriate for this type of setting and that the answers were evasive and fanciful.

On the basis of the census, a second survey was designed to follow up on those aspects of the questions on *lāg* and marriage support that had seemed to be more clearly established in sanctioned and public forms. It asked three groups of questions. First, it asked each respondent to list all those in his *melī*. Second, it asked for the names and place of residence of all those connected through marriage gifts (*niūndā*) in the last three years, and whether they had given or received them. Finally, it asked for all of the *lāgis* of the family, and their house numbers. This survey was administered by Naranjan and Netar Singh to ninety-three village families, all that would cooperate. Since this was more than enough of a sample to determine the degree of reciprocity, and reciprocity was itself the main sanction supporting the claims, the small number of missing families was not a serious problem.

A third survey was designed to get at party relations in a manner more appropriate than the questions about friendship in the census supplement. Since parties were defined as secret, we came up with a procedure whereby Naranjan and Netar, who were then each associated with a different party group in the village, would each ask a sample of men in their group who, in their opinion, the people in the two village alignments were. (For reasons which will be given in Chapter 8, there would have been no logical point in asking what group they were with personally.) The answers, however, had to be matched against my own records of those who attended the public rituals and legal conflicts that were normally considered party dominated and that represented the sanctioned application of the ideas. People in the same party supported each other's public rituals and cases; those in opposed parties would boycott each other's rituals and either failed to witness for them or would appear in cases against them. The result had to remain my personal view. There was by the definition of parties themselves no public way to confirm it, although there were actually no major disagreements among any of the sources I combined.

These three major surveys covered the major goods other than land that each person would control (houses, cattle, tools) and the people he would be directly associated with through relations of kinship, politics, and religion. Information on landownership was obtained from the official land records. No survey of my own design could have come close to being adequate.

At the time of my first study, it was commonly said among South Asianists that land records were not accurate, largely because of the "corruption" of the local revenue clerks. An exception to this was Oscar Lewis's description of the local records, which recommended their use, although he did not describe the institutional safeguards that made them more than *merely* reliable (1958:329–347). It is true enough, as Lewis himself noted, that local revenue clerks are typically grossly underpaid and that their salaries are a very small portion of the value of the land whose transfers they record each year. They sometimes receive gifts and tips and are legally entitled to fees for copying portions of the records—fees which villagers may construe as demanded bribes. Yet it is quite wrong to infer from this that the records themselves are unreliable, for two major reasons.

The first is that the village record is only one of three copies, and not the major one. It is considered a working copy. It is impossible for the local clerk to insert an error without discovery, which could cost him his job. (The actual method of recording a change in the record will be described in Chapter 5.) The second point is more basic. It is that the record (not the one in the care of the revenue clerk, but the one in the administrative headquarters) does not merely describe reality, it establishes it just as the sanctioned applications of the ideas of kinship or religion do. It is the system of title deeds for the village landowners and landholders. The revenue record carries a presumption of truth in a court of law. Villagers know it and act accordingly. In addition to the revenue record (Record of Rights) itself, I obtained a copy of the genealogy of landowners that went with it, a copy of the associated field map, and a copy of the genealogy from the first set of records made up in the village, in 1852.

The 1978 study repeated the general surveys of 1965 with two major exceptions. With no reason to look at weddings (and no time to wait for the weddings that would allow the reports to be checked) the surveys pertaining to *lāgi* and *melī* relations were left off the census. There would also be little point in repeating the more formal survey of party allegiance. I had been away from the village for too long to broach such a delicate topic so quickly and would have to pick up what information I could in passing. I planned to spend about one month working intensively in the village. Then, after my wife arrived from the United States, I would work mainly outside the village collecting background information on laws, on relevant institutions and institutional changes, and on comparable developments in at least one neighboring state.

Within this tight schedule, it was necessary to concentrate first of all on replicating the census, with additional information on tools and if possible more precise and consistent information on housing. The first census had taken more than three weeks to complete, and even then there were twenty houses I could not locate without the help of Netar, Naranjan, and Alekh Ram, the village postman. I knew they were somewhere because the official census in 1960 had placed a sequentially numbered plaque on the front door of the main living unit of each village family. When I discovered this and found the identification of family dwelling units to be reliable, I used them as my own household identification numbers. They had shown the gaps.

In 1978 it became clear on the first day that the village was larger, so it was reasonable to worry about the resurvey. At the suggestion of Surindar, Netar, Naranjan, and Naranjan's brother Mohan Singh, I decided not to go house to house. Mohan Singh was a teacher and had conducted the official 1970 census. He pointed out that the former system of numbering had been abandoned; the new system was to number every housing unit. Since there were often several to a family, the sequence would be no guide, and going house to house would be even more difficult than it had been previously.

The alternative was what we called the *chauk* system. A *chauk* is any public open area where people gather. There were four such places in the village. I had brought xerographic copies of the 1965 forms (just in case) and had new blank forms made up in Chandigarh. The method was to take these and go sit in one of the *chauks* with Netar, Surindar, Naranjan, Mohan Singh, or anyone else who understood what was intended. A crowd soon gathered closely around. It was then a simple matter of finding the forms to match the family represented by someone in the group (this is why the guide was needed), pulling out the appropriate old form, and filling in a new one—being sure to ask about every person previously listed. Except for a few times when I lost my concentration and forgot a question or two due to heat, fatigue, and crowding, everything went exceptionally rapidly and smoothly—so smoothly that the entire census was completed in three (long) days. This left more than enough time to return to Chandigarh and (with the help of my hand electronic calculator—a marvel unavailable anywhere in 1964) quickly make rough tallies of the changes that could then be brought back for discussion in the village.

Villagers themselves did not, of course, have a precise idea of such things as the changes in family sizes for the village as a whole or for each caste group, or the changes in cattle and tool ownership, and the information was interesting just in the way information on the general frameworks of ideas was not. Everyone knew "what everyone knows," but everyone did not by any means know what precise commitments and interests had been formed with the elements they defined. This was information on the working context for all future developments, and there was much discussion of how it had come about and what it would lead to.

THE OUTLINES OF CHANGE

The general information systems provide each villager with the intellectual tools for making decisions and calculating the relative value of various courses of action. At the same time, the commitments and interests each person has defined with his own particular successful imputations dictate what the decisions will be about and how successfully they can be implemented through recurrent patterns of cooperation and competition. Decisions in the context of commitments make up long-term social and economic strategies, and the interaction among strategies ultimately is what makes up the village pattern of development.

In many respects, the way individuals form such strategies is the most interesting part of this study, for it is here that character and talent reveal themselves. At the same time, in spite of the common idea that what is individual is highly "particular" and not susceptible of scientific analysis, it is here that we actually find the greatest commonalities across communities. We see beyond the different cultural tools that people use and into the common problems of using whatever tools one has; beyond the stereotyped ideas about people current in a community and into the way individuals handle the common problem of judging those around them with ideas that are never quite adequate.

Since the 1978 study was to focus on social and economic change, it seemed logical that the most relevant information on strategies would be found in the histories of the major changes in the village: the improvements to the gurudwara, the acquisition of the new school, and so forth. These were supplemented with a sample of personal histories, mainly from those directly involved in the major changes and from a few others I had known especially well during the first study.

The description of the changes in the information systems, the changing commitments and interests, and the shifting strategies that brought them about and reflect them should provide a systematic and complete account of the Green Revolution in Shahidpur. It was clear from the first moments of my visit, however, that the village did not initiate or sustain these changes in isolation, and it would be wrong to write in retrospect as if it did. There were "outside" events of crucial importance, and these will be indicated as we proceed.

The following three chapters will describe the many changes in the overall pattern of crops and the balance of animal populations that make up the ecology; in the pattern of household and corporate group cooperation and competition that makes up the system of management (division of labor); and in the marketing practices and the interplay of market exchange and price relationships that make up the economy.

There has been no major change in the system of kinship ideas, although there have been changes in the way some of the ideas are used. The most interesting story here is the way a governmental effort to impose change has been resisted. This will be given in Chapter 6. The changes in the use of certain important religious ideas and the idea of parties will be described in Chapters 7 and 8, respectively. Although no effort will be made to separate topics into chapters artificially, and several events will be seen to recur in new aspects in many chapters, Chapter 9 will provide some more intensive examples of the way various separate contexts are interwoven in individual lives, and in particular complex events.

Chapter 10 will conclude with a discussion of underlying motivations, concentrating on the idea of "rationality" and its use in analysis and policy design. Rationality is often treated as an exogenous and independent variable in culture change that might be discussed apart from any particular cultural system. But in reality it is nothing more than following certain rules, or rather "metarules," for decision making that are implicit in the information systems themselves as they are used. Indeed, they are implicit in any information system, in principle. It may seem impossible that there could be such transcultural rules implicit in the nature of information systems in general rather than any one or few systems in particular, but once one understands such systems it is easy enough to see. This will bring the theoretical wheel full circle, from the "largest" systems to the most intimate aspects of their use, and from the aspects of village life most strange to us to the aspects most utterly familiar. We will see thereby how the changes in this particular village in this particular state at this particular and unusual point in time reflect processes common everywhere, and thus finally how these results can be obtained elsewhere.

3

CHAPTER

The Village Ecology

An ecological system is a set of interdependent biological populations in an environment. There is no single village term that corresponds directly to our word "ecosystem," but the idea is present nevertheless. It is often expressed in a special notion of "the land" and what comes from it and must go back to it: the crops, cattle, and people.

In 1965 there was a clearly established assumption that the village ecosystem was self-sustaining within the physical confines of the village boundaries. Outside inputs were useful but not absolutely necessary. The many links in the basic cycles from the land through the crops to the cattle and human populations and back to the land were carefully maintained and little was left to chance—although people did not go to the length described by Srinivas for his village in Mysore. There, apparently, each family was careful to defecate only on its own lands and asked its guests to do the same (Srinivas 1976:16). In Panjab, modesty and convenience overrode ecological frugality.

By 1978 the quality of self-containment and autonomy had disappeared. This was a major change and was tied to further changes in the general orientation of villagers toward external institutions of all sorts.

The main separable elements of the system are the land and water supplies, the plant species, the animal species, and the various tools and shelters that permit them to be integrated with one another. From a village point of view, fertility is the "strength" of the land and should logically be discussed in the first section below. However, it will be more convenient and cause no serious distortion if I follow the Western practice here of describing it under its own heading.

LAND AND WATER

It may seem that the way a community thinks of land and water could not change very drastically in as short a time as twelve years, but it did. One measure of this is that in 1965 it was logical to consider land and water as two separate topics. For 1978 it was necessary to combine them.

The soil of Shahidpur is a uniform brown loam, neutral to slightly basic. There are no rocks and no bedrock. Its clay content is high enough to make pots and bricks. It holds water well but is susceptible to flooding. When wet soil dries, a crust can form that will choke off the shoots of some plants, including wheat. If wet soil is agitated, a hardpan results as the fine particles of clay settle out and form an impermeable layer at the base of the agitated zone. This too can choke off the roots of some crops, including wheat. The soil can be dug easily with light hand implements if its moisture is neither too high nor too low. Its nearly dead-level grade facilitates irrigation but augments flooding during the rains.

Land is broadly divided into agricultural and nonagricultural. There are numerous subdivisions of each, both official (in the land records) and informal. Nonagricultural land is subdivided into house sites and various types of use areas: roads, well sites, field channels, wasteland, and public areas of several kinds. In the legal records, agricultural land is divided into four major types according to its water supply: rainfall, well land, canal land, and well-canal land. These are increasingly productive in the order listed. Informally, the villagers in 1965 classified the land for crop planning into three subdivisions: poor, average, and best. The basis was "strength," which in turn was seen mainly as a function of water supply and distance from the village housing area. The land nearest the village had been farmed the longest, had the most abundant water (because wells had been built from the populated center outward), received the most human manure, and was easiest to supply with large amounts of compost (which was prepared in pits near the houses). Thus it was "strongest."

The land farthest away had been farmed the least time (some had been brought into cultivation only three years before my first study, according to a farmer who pointed out such a portion when I was walking by his field one day), had no wells, received the least human waste, and was most difficult to supply with compost. Hence it was "poor," the "weakest." Land in the middle, with wells, was "average." The exception to this general scheme was land along the canal, which was "best" even if it was far from the center. The canal had been completed in 1962. Crops were grown in different proportions on the different grades, the methods of cultivation differed, and the yields also differed.

In 1978 the legal subdivisions were the same, but when I asked about strength, I was told there was only best and average. There was no more poor land. The far land had not moved near, but wells had been built everywhere and chemical fertilizer more than compensated for the differential cost of moving the less-potent compost. There were still differences in yield, but the differences in crop patterns had evidently been either greatly reduced or eliminated. This, however, gets a little ahead of the story.

The official system of land measurement in India is now metric. Villagers also use the English system and several indigenous systems. For areal measurement, the basic unit of the most important indigenous system is the *bīghā*, which is divided into 20 *biswās*. There are 11.86 *bīghās* to the hectare, or 4.8 to the acre. Continued use of the *bīghā* to demarcate areas of land to rent, sell, buy, plant, water, and hire labor for is not simple conservatism. One *bīghā* is about the area one man can reap, weed, or hoe in a day; thus stated rates for paying people for these operations per day are also rates per *bīghā*. Actual planted fields usually range in size from one to about four *bīghās* (hardly ever an acre and never a hectare). Villagers have no difficulty using all these measures and often mix them (a practice that has caused me considerable confusion from time to time). In this study I will use mainly metric units, but occasionally will return to *bīghās* to show the connections between various conceptualizations. To follow local usage exactly would confuse far more than it would clarify

The agricultural lands are laid out roughly in the shape of a slipper along a north-south axis (see Figure 3.1). The Bhamian Canal Minor, a distribution channel of the Bhakra Mainline Canal that begins at a point about 22 km north of the village, cuts the land from about the northwest corner to a point in the middle of the western side. The main housing area, the *abādī* (from *abād*, meaning "population"), is in the center, surrounded by a square of roads called the *phirmī* (meaning "enclosure"). The roads leading out from the *phirmī* to the village borders go to neighboring villages.

In 1965 the total village land area was 178.26 hectares, of which 15.4 were nonagricultural. In 1978 the total was 178.33 hectares, of which 17 were nonagricultural. The slight increase in the total represents land purchased by local families in an area just outside the old village land boundary. The increase in nonagricultural land reflects the increase in house sites, well sites, and roads.

The traditional pattern of highly nucleated construction in Panjab villages

Figure 3.1 Map of Village Lands, 1978

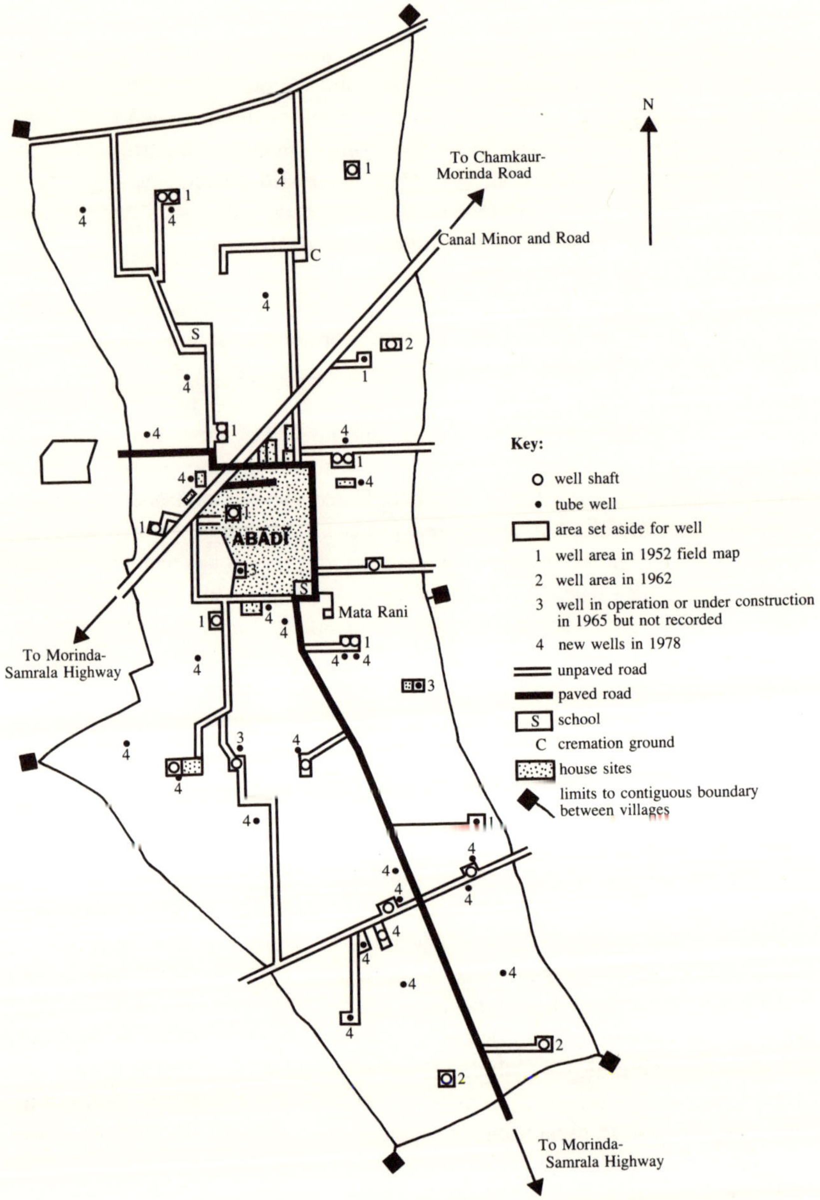

has two main functions: to provide a cooperative fortification and to avoid encroaching on the farmland. In 1965 all of the houses (and all but one building) were still within or on the *phirmī*, but their outer walls no longer formed a complete rampart and some houses had been built standing apart. By 1978 the breakdown had become much more obvious, as already indicated. In addition to the housing along the two new lanes directly north of the *phirmī*, five farmers had built new houses entirely away from the *abādī*, in their fields.

There were related changes in household architecture. In 1965 new houses were still being built predominantly on the old pattern: high outer walls with small windows or no windows, massive iron or wooden gates, and if possible an enclosed inner courtyard. In 1978 most new building was of a more urban type: compact houses with no inner courtyards but with large exterior windows, surrounded by fairly spacious yards with low walls or even incomplete walls. The decorations, too, were like those of city houses: painted walls, screens on windows, and even a mail slot in one new gatepost. The changes in house arrangement and architecture obviously reflected greatly increased willingness to rely on state law and police for personal security—just one of the ways village autonomy had broken down.

The same confidence in regional law enforcement could be seen in the increased use of mechanical pumps for drawing water in the fields. Such pumps were much more portable and more valuable than the old Persian wheels. Once removed, they would be almost impossible for villagers to recover on their own—especially in view of the improved roads. And not all pumps and motors were kept in locked brick houses. Some were in what Naranjan's brother called a "Panjabi pumpshed"—a heavy plastic sheet over a pump bolted to a stand-pipe at ground level. Farmers could not afford to replace many such pumps before the costs would become unbearable (there is no insurance).

In 1965 canal-watered land was 11.8 percent of the total. Well-canal land was 4.4 percent, well land was 61.8 percent, and rainfall land was 22 percent. In village categories, this meant that about 16 percent was "best," 62 percent was "average," and 22 percent was "poor." By 1978, according to the official record, canal land had increased to 14 percent, well-canal to 9.7 percent, and well land to 75.3 percent. Rainfall land had been reduced to 0.5 percent. Thus, roughly 26 percent was "best," and 74 percent was "average." Yet because of the new water-lifting technology, the change in the categories is only part of the story.

As Figure 3.1 indicates, there were twenty-six well shafts at the end of 1965, an increase of three from 1955/56. Between 1965 and 1974/75, when the record was redrawn, twenty-seven more wells were added while one went out of service, thus doubling the former total. At the same time, each new well produced about twice as much water as the old type. Hence the total water-lifting capacity in 1978 must have been about four times what it had been in 1965.

The reason for the spurt of well construction in the years just prior to my first study lay in a rise in the underground water table, an unintended consequence of the completion of the Bhakra Canal. The Bhakra was built as part of the reconstruction of the Indus Basin irrigation system after partition in 1947 (see Michel 1967 for an excellent overview). It begins at the Sutlej River above Ropar, in the lower Himalayas about 50 km north of the village and passes within 3 km to the east. Its distributary minors run well to the west of the village. All the canals have permeable bottoms and inject large amounts of water into the ground along their courses. Before the Bhakra system was opened, the water table in the village was, reportedly, at thirty meters. Wells were very expensive and they were built only at long intervals. The local water table started to rise immediately after the Bhakra Mainline Canal was completed in 1954 and rose even faster after the distribution minor came to the village in 1962. By 1965 it had reached ten meters. This reduced the cost of a well proportionally and made well building a very attractive way for an individual farmer to increase productivity. By 1978, the level was at seven meters and still rising. The mechanical pumpsets thus did not create the present situation by themselves, but accelerated a trend toward cheaper and more-abundant water supplies, obtained by personal investment, that was already underway.

In 1978 it was believed that the threat of crop loss due to drought had probably been eliminated, although no one was sure because the monsoon rains had not failed since 1965. In the summer of 1979 Surindar wrote that this theory would apparently be tested; the rains had failed. Subsequently, he said the crops had been excellent.

FERTILIZER AND FERTILITY

Fertility is spoken of as a manageable aspect of the "strength" of land, closely related to the water supply.

In 1978, as in 1965, the nutrients consumed by village crops came from four main sources: fallowing, crop rotation, compost, and chemical fertilizer. Between the two points in time, the use of fallowing had been reduced, reliance on crop rotation decreased somewhat, and the use of compost and chemical fertilizer increased greatly.

Chemical fertilizer usage varies according to the crop, the farmer's purpose in growing it, its rotational position, and compost availability. In 1965 three main chemical fertilizers were used: urea, CAN (Calcium Ammonium

Nitrate), and "American" (superphosphate). Superphosphate was imported from the United States and was expensive.

By 1978 two new concentrations of superphosphate had been added, along with zinc sulphate and muriate of potash. All were made in India. Muriate of potash was not currently in use. Farmers had recently tried it (following Punjab Agricultural University recommendations) and seen no improvement in yields.

When I first asked about quantities of fertilizer used in 1978, the questions were answered with disinterest. As I persisted, the reason for the disinterest was given: everyone followed the recommendations of the Punjab Agricultural University; I did not need to ask the quantities, I should just look them up. In 1965 the topic had been of much greater interest and answers indicated amounts anywhere from 20 to 100 percent lower than official recommendations. The reasons for the discrepancies at that time included lack of sufficient agricultural credit to support the increased risk that increased fertilizer usage would entail (in the event of drought) and lack of water (or credit for building wells). The reduced disagreement among farmers and between farmers and official recommendations indicates that these underlying earlier problems have been brought under control and that the official recommendations have come to reflect a much sounder grasp of village realities.

In 1978 official fertilizer recommendations were published before each season as part of a package program describing an entire recommended crop pattern. District recommendations were made for each major soil/climatic subregion of the state, given the expected weather and the results of the last season, so they were quite detailed and specific. The state government undertook to make balanced supplies of fertilizer and water available in rural areas according to the plan recommendations. This effort was somewhat less successful between 1972 and 1977 than before or after (Gill 1983). Block Development Officers received the recommendations for their areas and passed them down to Village Level Workers. The recommendations were also publicized by radio and in news magazines published by Punjab Agricultural University. Farmers took them seriously, although they still had to modify them according to their estimate of the performance of their own land, their own purposes, and the materials they might already have on hand.

To make comparisons, it is necessary to group together the various types of fertilizers under a uniform measure. Farmers do this using cost as an estimate of amount. In 1965 a standard allowance for sugar cane on the best land was Rs. 13 a *bīghā* (equal to Rs. 153 a hectare). For comparably high yields of wheat, maize, and cotton on land of the same quality, the allowances were about half that amount. Table 3.1 compares these allocations to their counterparts for 1978. In the table, the 1965 figures have been multiplied by 1.5 to cancel the effect of price increases. In 1965 the fertilizers

Table 3.1
Total Fertilizer Allocations by Major Crop, 1965 and 1978

Crop	1965 (Rs. per hectare × 1.5)	1978 (Rs. per hectare)
Cane	229.50	941.00
Cotton	114.75	———
Maize	114.75	547.00
Rice	———	470.00
Wheat	114.75	353.00

most commonly applied were urea and superphosphate, in a ratio of about 2:1. By 1978, the price of urea had increased by 40 percent, the price of superphosphate by 68 percent.

In 1965 fertilizer was generally purchased by farmers out of cash savings, although government *toccāvī* loans at 2.5 percent interest were available if farmers had the time to travel to Ropar, about 12 km northeast along the canal roads, for the paperwork and to make the repayments.

Beginning in 1968, the village cooperative society began to issue loans for fertilizer. The loans were in kind and the cooperative secretary arranged to bring the fertilizer to the village from nearby depots. The records of this society provide an indication of the way the changing allocations at the individual level add up for the village as a whole. There were several organizational and record-keeping changes between 1968 and 1978, so it isn't possible to obtain an absolutely complete and coherent set of accounts, but the records in the depot office in Morinda show the villagers' total fertilizer loans issued for 1969/70 (the first year I could locate) as amounting to Rs. 14,000. The village cooperative's records for 1974/75 (the first year of this record) show loans for Rs. 50,701. The same record shows the new loans for 1977/78 to be Rs. 70,505. Allowing for the noted increase in unit prices and for the probability that some farmers obtained fertilizer from noncooperative sources in the first year or two, there still seems no reasonable basis to doubt that these figures represent 1978 fertilizer usage easily 300 percent of the level of 1964/65.

There has been no change in the basic method of fertilizer application. It is drilled into the earth with a plow, either before or during sowing.

Compost complements chemical fertilizer. If there is sufficient nitrogen in the soil to support its decomposition, it improves soil texture and drainage and adds "strength." Compost is made from household and barn waste, kitchen ashes, and, increasingly, plant residues directly from the fields. Houses with cattle, especially those with cattle kept in interior barns, contract with Sweepers to collect manure that is not clean enough to be dried for fuel. Each day this is swept up along with other waste and dirt from the barn floor and from the street in front of the house (up to the middle of the lane) and carried out to a compost pit or heap maintained on the family's land, on common land near the *abādī*, or in pits near the canal. In 1965 little or no material was added to the compost directly from the fields. Everything was first either chopped for fodder or dried for fuel. These intermediate uses were too important. Even dried roots and weeds would be collected to burn. By 1978 fuel was evidently abundant and the coarser residues were no longer being chopped for fodder. More than half the material in the pits seemed to come directly from the fields.

An estimate of the amounts of compost used can be made from the area used for pits and heaps. In the winter of 1965, the heaps occupied an area about forty meters square near the *abādī*, and the pits lay along one side of the canal for a distance of about twenty meters. In June 1978, the area of heaps near the *abādī* had shrunk to about fifteen by thirty meters, but lines of pits extended down both sides of the canal for a total length of about two hundred meters. Allowing for seasonal variations in the amount in storage, it seems safe to say that compost usage must have increased at least threefold, corresponding closely to the increased use of chemical fertilizer.

The importance of rotating nitrogen-consuming crops such as grains (grasses) with nitrogen-fixing legumes was thoroughly appreciated in both 1965 and 1978. However, since the main improvements in yields came in crops that were heavy nitrogen consumers and the increased yields were accompanied by increased acreages planted, opportunities for such rotation were automatically reduced. Still no one argued that this reduction could be entirely compensated for by the appropriate application of chemical fertilizer and compost. Farmers said that a heavy nitrogen-consuming crop following one or more other such crops (like rice following wheat, or potatoes following a summer grain following a winter grain) would always yield less than the same crop, fertilized the same way, following a legume or fallow.

Fallowing is the most difficult source of soil nutrients to evaluate. Here, also, there was a reduction between 1965 and 1978. This occurred because the increased water produced an opportunity for more catch-crops between the main crops and because some of the new crops required longer growing seasons. Yet farmers in 1978 still insisted that if the land was allowed to "rest" for at least a month after a grain crop, it would be "stronger." Without this rest, the following grain yield would be measurably reduced. This, I was told, was exactly why I had seen so many cleared but unplowed fields when I first arrived. Farmers were waiting to the last possible moment, which would be just when the rains began, before breaking ground for the next crops. This is evidently more than mere folklore since nitrogen has been shown in controlled experiments to be fixed by free-soil bacteria of several genera, quite apart from bacteria in symbiotic relations with legumes, and these reactions can be quite productive under favorable conditions (Rajan and Rao 1978: Chapter 10). At present, however, there seems no way to put a verifiable quantitative value on the village statements.

In short, while there have been losses as well as gains, and the farmers do not believe that the losses are entirely compensated for, there is no doubt that overall the effective fertility of the land has been substantially upgraded.

CROPS

The new wells and increased nutrients have been associated with greatly increased yields, although the increases vary markedly by both crop and land type. Increased yields in turn have led to a new overall crop pattern.

The increase in yields of the main commercial and food crops are shown in Table 3.2. By far the most dramatic gain is in wheat grown on average land, which Table 3.1 shows has the lowest fertilizer expense. The smallest increase in yield—none—is for cane on the best land, for which the fertilizer expense is greatest. This is what is meant by fertilizer responsiveness, and it is obviously significant. Cane in 1965 required more fertilizer than any other crop and it still did in 1978. Wheat took more fertilizer in 1978 than in 1965, but it gave proportionally much more back. The yields, especially for wheat, are high even by United States standards.

For planning purposes, villagers classify crops both by what they require and by what they provide. What they provide comes under the headings of food, fiber, fodder, medicine, and, increasingly, money. What they require is stated in terms of the money, water, labor, and land they absorb. Farmers also consider the difficulty of growing each crop. Other things being equal, they would rather grow a crop that is less difficult.

In 1978 as in 1965, the crops were closely adapted to the local climatic cycle. In this area, the four annual reversals of monsoon winds produce six recognized seasons of almost exactly two months each. Traditionally the new year begins on Baisakhi Day, the first day of the month of Baisakh. Indian months (in the Vikram calendar and several others) go from full moon to full moon. Baisakhi Day is the day of the new moon that falls in the Julian month of April—generally April thirteenth. The first day of Baisakh is also the beginning of the "hot" season of light winds and increasing heat which lasts until mid-June. In mid-June the "rains" begin with the arrival of the water-bearing monsoon winds from the southeast. About two-thirds of the yearly average of eighty-three centimeters of rain falls in this season. In mid-August, the monsoon winds slacken and a second "hot," dry season be-

Table 3.2

Yields of Major Crops, 1965 and 1978 (Quintals per Hectare)

Crop	*Best land*		*Average land*		*Poor land*	
	1965	*1978*	*1965*	*1978*	*1965*	*1978*
Cane	1,000.0	1,000.0	500.0	824.0	NG	NG
Cotton (American)	22.0	NG	NG	NG	NG	NG
Cotton (*desī*)	22.0	24.0	NG	20.0	NG	NG
Maize	26.0–40.0[a]	35.0	20.0	24.0	6.0	12.0[b]
Rice (*bāsmatī*)	NG	23.5	NG	17.7	NG	NG
Rice (*Jāya*)	NG	74.0	NG	47.0	NG	NG
Wheat	42.0	47.0	26.0	41.0	13.0	24.0[b]

Note: NG = not grown.

[a]This high figure was obtained with a hybrid variety in an experimental plot set aside by Indar Singh. The yield was very unusual even though yields that year were very good on irrigated land in Panjab as a whole.

[b]This means the poorest grade of well-irrigated land, not rainfall land.

gins. In mid-October the winds shift again. Blowing from the north, they bring the "cold" season. This is sometimes called "fall," in English, for it is when most trees lose their leaves. They immediately grow back, however, in the season of "rains" or "cold rains," beginning in mid-December, followed by a final "cold" season before the new year.

High temperatures in the first hot season occasionally reach 46°C (115°F), although temperatures in the range of 40°C are most common. In the cold season, lows usually reach just above freezing, producing heavy morning fogs and occasionally a light touch of frost. Daytime highs in the cold season range into the upper twenties. Thus for a good eight months of the year, from October through May, the weather is comfortable and pleasant.

All crops need more water at some points in their growth cycles than at others. Most crops can also be damaged by too much water at some times and most must be dry when harvested. Hence in 1965 the farmers' strategy was to plant the crops in such a sequence that their period of maximum water need occurred at the time of the maximum rains and their period of maturity fell in the following dry, hot season. With the great increase in the portion of the total water supply under human control, part of the need for this precise timing was eliminated, permitting much greater cropping freedom.

Traditionally, the two harvest seasons divide the year into two crop periods. The first crops of the year are the summer crops, called *kharīf*, harvested in the September hot season. The crops harvested in the spring are *rābī*. Table 3.3 shows the *kharīf* crop patterns of 1965 and 1978. The changes are quite substantial. The largest increases have been in maize and rice. Before the Green Revolution, rice was not considered a traditional Panjab crop. In 1960/61 it occupied only 227,000 of the 5,038,000 hectares that were to become Panjab State after the 1967 division (Agriculture Department, Panjab, 1964). There were still only 292,000 hectares in 1965/66, but by 1974/75 the area had grown to 596,000 (The Economic Advisor to the Government, Punjab, 1974) and to 1,052,000 in 1978/79 (Gill 1980). Over the same period, but with an especially sharp jump from 1968 to 1979, average yields per hectare nearly tripled.

Rice was first sown in Shahidpur after electrification, in 1970. Its area greatly increased in 1977. Electric fees are charged at a flat monthly rate based on the size of motor or type of installation. In 1970 the rate was a very nominal Rs. 40 a month for a 5 HP electric motor. In 1975 it went up to Rs. 95 a month. In 1977 it was reduced to Rs. 65. The fixed rates encouraged farmers to find ways to use as much water as possible per month in order to make the highest return per well per unit of *time*, rather than per unit of *water*. Rice was an obvious answer. Remaining in water the entire time it

Table 3.3
Kharīf *(Summer) Crop Pattern, 1965 and 1978*

	Percent of total area under each crop	
Crop	*1965*	*1978*
Cane	31	27
Chilis	1	1
Cotton (American)	9	0
Cotton (*desī*)	8	1
Fodder	12	14
Maize	21	40
Mung and mash	2	1
Rice	0	10
Other	14	6
Failed	2	0
Total	100	100

Source: Farmers' estimates.

grows, it uses far more water than any other crop. But it yields a more reliable financial return per hectare than any other *kharīf* crop, and requires little weeding. Two improved varieties were grown in 1978. One was a very fine quality, tall-statured *bāsmatī* (long-grained indigenous type). The other was a less-favored, short-statured variety called *Jāya*. The *Jāya* yielded about twice as much grain per hectare but the market price was just about half, so the financial return per hectare was the same. Rice is grown mainly for market; it is seldom eaten in the village.

Rice posed new and rather paradoxical problems for soil management. Although for many crops the soil of Shahidpur has a rather high clay content, for rice it is rather low; water in a flooded field will soak into the ground relatively quickly. To limit this loss, it is necessary to "puddle" the fields, that is, to deliberately create the hardpan that is normally avoided at all costs. This hardpan must be broken up after each rice crop by exceptionally deep plowing. The ordinary plow cannot do this well (certainly not if it is used to create the hardpan in the first place, because the hardpan will always be just *below* the depth of the cut that made it). Accordingly, two versions of a deeper-cutting, furrow-turning plow have been included in the official package program materials (Agriculture Department, Panjab, 1964). I did not know this in time to include separate plow types in the tool inventory of my census and did not stay long enough to see the harvest, but from the firm reports that the yield of a crop after rice remained substantially less than after other crops I infer that the problem was not solved.

Maize was a major food and fodder crop in both 1965 and 1978. The grain is eaten as flour, not fresh. The stalks and leaves are dried and chopped as fodder and corncobs are an important cooking fuel. Maize *roTī* is the preferred winter staple. In 1965 it was a standard practice to hold back enough of the wheat harvest (in spring) to supply the family's needs for the entire forthcoming year and then wait to see the maize yield. When the maize was harvested, an amount of wheat equal to the harvested maize would be released for sale and the maize would be kept. Only rarely was there enough maize to sell directly, because the importance of maize as fodder overrode its value as food. Accordingly, it was planted more densely and given less water than if the intent had been to maximize grain yield. It was often grown on rainfall land.

In 1978 maize was grown only on irrigated land because that was the only kind of land there was. There was adequate water and fertilization for high grain production, while plenty of fodder was available from other sources. Hence the strategy shifted to emphasize grain production. Enough maize was grown to exceed local dietary requirements, so it could be marketed directly. There were also improved varieties.

In 1965 the main variety of maize grown had been a local stable race, simply called *desī* (meaning "local," from *des*, meaning "country" or "region"). Each year farmers would take their seed from the earliest and largest ears of their own crops. It was hardy, drought resistant and fast growing, but grain yields were low.

When we first talked about maize in the village, just after the fall harvest of 1964, I was asked what proportion of plants in the United States had ears (not how many ears there were per plant). I asked why the question was put that way and was told that in the village one ear on three plants was considered good. Yet even then a few farmers were experimenting with more-productive varieties on better land, using commercial seed. After 1965 the *desī* type was largely replaced. The first new variety was Composite Vijay, which had been created by genetic transfer. By 1978 this had been supplanted in turn by a hybrid called Ganga #5. Composite Vijay had the advantage of producing consistent seed, so the farmers were able to take new seed from their own crops for a few seasons instead of making fresh purchases every year. With the hybrid, farmers could not use seed from their own harvests at all, so the old reproductive cycle was broken—or rather extended to include the workings of the Plant Breeding Department of Panjab Agricultural University. Some *desī* maize was grown if farmers could not plant the hybrid in time to reach maturity before the winter rains began (perhaps because of a preceding between-season crop after wheat), or if they wanted to clear the land early in the autumn to insert a following between-season crop, usually potatoes. The *desī* maize requires about 70 days to reach maturity; Ganga #5 requires 100 to 110 days.

Table 3.3 shows that the shift to more maize and rice has been mainly a shift away from cotton. Cotton is a major cash crop, and also produces fiber and fuel. The fiber was used locally to make homespun cloth, yarn for weaving cots and the like, and as stuffing for quilts. For this, the farmer selected the earliest-ripening and best bolls of his crop. They were ginned in the village and the seed was used for the next crop. Surplus seed might be fed to animals, but it was not a significant fodder. Dry stalks were used as fuel, but were not favored. For cash, cotton was sold unginned to itinerant buyers. Cotton continued to yield good cash returns in 1978. One reason for the shift away from it was, as already suggested, the fact that clothing and cloth were almost all ready-made. In addition, the changes in the village ecological base eliminated some of its former noneconomic advantages.

While cotton requires "strong" soil, it does not require, nor does it benefit from, much water. Also, the time of harvesting is especially critical for cotton. The weather must be dry the entire time the bolls are open before being picked. Rain causes fungus diseases that destroy the quality of the fiber. The rigidity of these requirements led to two different types being grown in 1965. The *desī* variety benefited from more water and matured earlier. The American variety was more drought resistant and matured later. In an early monsoon, conditions would favor the *desī*. The American variety would do better if the rains were delayed or continued longer than usual. By 1978, water conservation was no longer a consideration, while the long maturation time of the American type had become a relative disadvantage in view of the increased time to harvest of the newer crops and the possibility of a vegetable intercrop in the erstwhile dry season. Thus the American variety was completely eliminated, while the area under *desī* cotton was sharply reduced.

"Mung" and "mash" are legumes. They are grown as food (the beans), fodder (the greens), and as a means of strengthening the soil between grain crops. "Other" crops indicated in Table 3.3 include chickpeas and vegetables of various kinds, grown both commercially and for local consumption. The fodder indicated in the table was primarily pearl millet (*bājrā*), which in 1978 was also being grown for spring as a between-season crop harvested in June. *Bājrā* is drought resistant and likes warm weather, but is air pollinated and does not set seed well if there is heavy rain when its heads develop. It is stored dry and chopped for mixing with wheat or rice straw, as well as fresh green fodder.

The winter (*rābī*) crop pattern is given in Table 3.4. Here the main point is the increasing domination of wheat. Wheat grain is a major market crop and dietary staple. Wheat straw is the main base of dry fodder, into which the more nourishing dry and fresh components are mixed as required by each animal. The 1965 figures for wheat areas reflect the practices of the

Table 3.4
Rābī *(Winter) Crop Pattern, 1965 and 1978*

Crop	*Percent of total area under each crop*	
	1965	*1978*
Barley	1	1
Barley and gram	1	0
Cane	31	26
Fodder	10	1
Gram (chickpeas)	6	1
Lentils	1	1
Oil seeds	2	1
Wheat	21	69
Wheat and gram	24	0
Other	2	0
Failed	1	0
Total	100	100

Source: Farmers' estimates.
Note: Cane is planted in one summer and harvested repeatedly for the next three winters.

region immediately around Shahidpur, not just the village itself. They are based upon official records, discussed in the village and adjusted in the light of proportions of crops paced off on village roads. (Actual physical measurement would have been almost prohibitively time consuming and diplomatically unwise at that point, since it would have made my work look very much like that of government agents associated with tax-collection.)

Despite their convenience, however, the regional data distort one important point. In Shahidpur itself, much less wheat was grown intermixed with gram (chickpeas) than in the neighboring villages. Most wheat and gram were grown separately, each in their own fields. The yield was slightly higher with mixed sowing on average or poor (rainfall) land, and the risk of total loss in a drought was much reduced since gram is highly drought resistant. On better land, however, the harvest and threshing costs are much higher for the mixed fields, and even then Shahidpur (because it was older) had substantially more irrigated land than the surrounding villages.

It has been argued that there has not been a Green Revolution, but only a "Wheat Revolution" (Srinivasan 1971). That is, there has not been an overall change but only the introduction of one really successful new crop. It should already be clear that this is not correct. Yet wheat has indeed become exceptionally important.

The story of the change in wheat farming in Shahidpur, in one sense, began in Mexico, in the International Centre for the Improvement of Maize and Wheat, a joint venture of the Rockefeller Foundation and the Mexican Government. In 1963 seed for new strains of wheat were received in India from Dr. N. E. Borlaug, the head of the Centre. The seed was distributed to plant breeders in various places, including the Panjab Agricultural University. Scientists at the University selected two promising strains and began multiplying the seed. Taking advantage of the fact that ecological zones of the state ran virtually from the deserts of Rajasthan to the high Himalayas, wheat was grown in the plains during the usual winter cropping season and then, immediately in the next summer, in mountain areas that had comparable climates. Thus two generations were produced each year instead of one (Randhawa 1975:19–21).

Some of the new seed was distributed in Panjab in 1966 for use on a demonstration basis. Statewide, there was enough for 100,000 acres in 1966/67, mainly in Ludhiana district. (Shahidpur was not in Ludhiana, but was just across the district boundary to the east.) By 1970/71, there were 500,000 acres (Randhawa 1975:20–21).

Both initial varieties were two-gene dwarfs. "Dwarf" means that while the grain size is normal, the straw of the plant is short. "Two-gene" means that a particular variety has been created by transferring two specific genes to a formerly different type. The new genes have been incorporated into the genetic structure of the new plant and are reproduced in cell division along with the original genetic material. The reproduction is however much less perfect than it is with a stable natural race. There is unpredictable variation in succeeding generations and a broad tendency for the qualities created by the transferred genes to deteriorate. Experimentation for still better and more stable varieties has continued, while at the same time an organization was built to provide the new seed to farmers not once but repeatedly. This is what became the Seed Store of Panjab Agricultural University.

Farmers in 1978 said they bought about one kilo of seed and the appropriate fertilizer for a seedplot. The crop produced was used as seed for the first field crop. Yields are highest in the first year and drop successively after that. Farmers reported that yields on good land were about forty-seven quintals per hectare in the first year, forty-two in the second, and thirty-six in the third. At the end of four years, they drop to the level of the better *desī* types, so fresh seed has to be obtained at the Seed Store and the cycle started over.

By 1978, continued experimentation had resulted in three major classes of wheats. There were improved *desī* types, created by selection from indigenous varieties that were tested and evaluated under experimental conditions that duplicated actual farm practices (rather than the older types of experiments that were intended more to set up demonstrations of what farmers *should* be doing). There were two-gene Mexican dwarfs of several types and new semidwarfs created by transferring genetic material from the Mexican dwarfs to the taller *desī* types.

The varieties differ in grain quality, fertilizer responsiveness, drought resistance, disease resistance, grain yield, straw strength, susceptibility to flooding and lodging (when, in wet soil, a plant loses purchase, falls over, and is unable to right itself), and fodder production. One of the reasons for the continuing effort to seek newer forms is to obtain a range that will allow farmers more latitude in planning. Another is that the first varieties have shown gradually increasing susceptibility to local diseases.

Farmers in Shahidpur were fully aware of the difficulties that new varieties entailed. They were quite willing, on balance, to take the risk of abandoning their old seed strains and thus showed a confidence in governmental institutions I would not have thought possible in 1965.

The first of the new wheats came to Shahidpur in the winter of 1968/69, a dwarf variety called simply Mexican (by those telling me the history in 1978). It gave a good yield but was subject to lodging. In 1970/71 Sonora 64 was introduced. It avoided lodging but did not produce enough fodder. By the following year, four more varieties were being used, including some of the semidwarfs and a late variety that could be sown in the first half of November and harvested after Baisakhi Day. The late variety was Dara 21; the others were Sonalika or S 308 (with superior grain), Kalyan Sona, and PV 357. By 1978, village farmers were growing mainly PV 357 and Sonalika, and two still-newer varieties: NL 711 and 2011. As with rice, the yields and quality so nearly balance each other that the monetary return per hectare is substantially the same for all the varieties, so the reasons for a farmer to select one or the other generally rest on ecological, rather than economic, considerations.

Among the crops grown less extensively than wheat, gram (chickpeas), mung, mash, and lentils, all appear to be produced in direct response to local dietary requirements and cannot be reduced further. Barley was grown in 1965 as a fodder and as an emergency grain for human diet because of its drought resistance. It was grown on rainfall land, usually together with chickpeas. In 1978, with plenty of water and no fodder shortage, there was no further need for it, so it was eliminated.

Oil seeds grown in the winter are mainly mustard. In 1965 mustard was often seeded in the same fields as barsheen, a type of clover. The mustard grew faster, and while it grew the barsheen provided it with nitrogen. Leaves of young mustard were picked as a green vegetable. When the mustard stalks and pods matured and were cut for pressing, the barsheen grew in its place. Oil was important in the diet. The mustard residues were used as a fodder supplement. The barsheen was cut daily as a green fodder. These usages continued in 1978 but mustard oil had also come to be used as a general balm and skin lotion; in 1965 it had been too valuable for such applications.

Although the connection may seem obscure, the need for oil seeds has been reduced by the greater fodder production. Contrary to widespread belief, in a vegetarian diet it is not so much protein that is in conspicuously short supply but fats, needed as a bodily fuel. In Panjab the preferred fat for food and food preparation has always been butter (or rather butterfat, called ghee), but in 1965 the chronic fodder shortage made this scarce. Mustard oil and sesame oil supplemented the ghee supply, and a great deal of peanut oil was purchased from outside the village. In 1978, however, ghee was plentiful. I am sure this is one of the reasons farmers felt free to introduce vegetables such as squash in place of the oil seeds—so that in consequence they could eat a vegetable other than the greens produced as a concomitant of the oil.

Sugar cane is a crucial component of the local diet, providing iron (scarce in the absence of meat) and energy (scarce because fats were scarce). It is also a major cash crop and produces some fiber (for rope) and fuel (the bagasse of each year's squeezing is saved in a stack and used to boil down next year's juices), but no fodder. Between 1965 and 1978, cane area declined by about 19 percent. Since production per hectare was unchanged, the decline in total output was also 19 percent.

Cane is sown early in the summer and grows through the hot weather. It becomes dormant when the weather cools but retains its sugar content. It is cut throughout the cold season and winter rainy season and is either sold in bulk or squeezed and boiled down locally. The plants are allowed to grow back again from the same roots the next summer. This ratooning is repeated for up to four years. It is said that the sugar yield continues to improve throughout this time, but the pests build up as well. At the end of the fourth year, the roots must be plowed up and the cycle started again on land where cane has not recently been grown.

In 1965 cane was the single most-remunerative crop per unit of land, but it required the most water and labor. By 1978, all the major crops required about the same labor as cane, but cane was less profitable over a full year than a maize-wheat rotation and about the same as a rice-wheat rotation (which provided fodder and allowed for an intercrop), although farmers said this could change overnight with a rise in the cane price at the government-owned sugar refinery in Morinda. The government had announced that sugar prices would be decontrolled after August 18, and the Morinda mill had announced in July that its capacity would be doubled as of November 14, after a major refitting. The dietary importance of cane remained unchanged.

These field crops were grown in rather large plots of from .25 *bīghās* to several *bīghās*. Apart from them, many village families (even those without commercial farms) had small garden plots of a hundred square meters or less where they grew a long list of vegetables for daily consumption: cabbage, cauliflower, carrots, chilis, coriander, eggplant, garlic, onions, melons, peas, potatoes, tomatoes, squash, and occasionally a little cannabis.

Other important cultigens in the village ecology include trees and several species of cactus. The main trees are kikkar (*Acacia arabica*), neem (*Azadirachta indica*), and mango. Trees provide important shade at well sites and work areas in the summer, and are used as a source for all manner of implements—tool handles and the like—for which wood is cut off as needed. Twigs of kikkar are the indigenous toothbrush and infusions have medicinal uses. The thorny branches are used to make fencing to keep wandering buffaloes out of fields and off canal banks. Neem leaves are an important green fodder. Mangos are virtually the only fruit regularly eaten. The cacti are used as living barbed-wire fences, transplanted wherever needed. There are also a few semiwild plants, most notably a native tobacco, allowed to grow and used for several purposes.

The total village production of any set of crops can be estimated from the yield of each crop on each type of land multiplied by the amount of land of that type. My previous estimates (Leaf 1972:387) put the sum of yields for the major grains (maize and wheat) in 1965 at about 227 metric tonnes. Assuming an even distribution of crops over all land types in 1978, the outrun in 1978 for the major grains (maize, wheat, and rice) would have been about 731 tonnes. That is, the 1978 production was just about 322 percent of the level of 1965. This does not translate directly into a comparable net increase in total food production because some of it was gained at the cost of legumes and sugar cane, as noted, but even when the losses are allowed for, it is clear that the net gain in food production must be very substantial.

LAND USE STRATEGIES

Although total production depends on the way crops are associated with specific types of land and I do not have comprehensive direct information on these associations, there is good indirect evidence in the land records that the pattern has changed substantially over time. More exactly, in 1965 there was a definite pattern; in 1978 it had broken down. The association of crops with land is related to the way land of each type is associated with land of the other types, and this can be measured precisely. In 1965 there was a strong tendency for those who obtained income from well land also to obtain income from rainfall land (Spearman's rho linear correlation of .64). This correlation was strictly for land as used; the correlation between well land

and rainfall land that each farmer owned was less, .53. There was also a moderate tendency (correlation of .33) for well land to be used in combination with canal land and well-canal land (.40). Apart from their use with well land, there were no tendencies for canal land or well-canal land to be used with any other type.

These associations mainly reflected the underlying chronic fodder shortage. Well land required the most draft power, hence it required the most fodder consumption for each unit of area farmed. A farmer could grow fodder on the well land itself, but there were also two other possibilities. Rainfall land, which was cheap, could be used to grow such drought-resistant fodder crops as barley mixed with gram or maize. This practice would free well land for crops like cotton and cane—profitable but not productive of fodder—which would not grow on rainfall land at all. Canal land or well-canal land was also available for rent or lease, partly because it was bought as an investment by farmers moving out of the village. It too could be used for fodder production. Accordingly, one would see on canal land not only remunerative cash crops such as cane, but also pearl millet, which, while drought resistant, also does very well when watered and fertilized heavily. Canal land was also often used in very small parcels by families not engaged in commercial farming. It provided fodder for household milk production and lentils, grains, and cane, which required no supporting production from land of other types. Well-canal land was the most productive type of all, and those who owned and used it had no need to supplement its harvest by fodder production on well land (which would be irrational), canal land (which in effect they already had), or rainfall land.

The comparable tabulations for 1978 show no strong tendencies for land of any one type to be used in conjunction with land of any other particular type, but rather a moderate to weak tendency for land of *any* type to be associated with land of *any* other type. All correlations (except that between well-canal land and rainfall land) were about equal, and between .34 and .38. The correlation between the amounts of well-canal land and canal land used and rainfall land were weak, .18 and .24 respectively. The breakdown of the 1965 pattern directly reflects the cropping freedom gained from the new technologies. The drought-resistant crops are not grown at all, because their yields are too low, and all remaining crops are grown about equally on all remaining land types.

The especially low correlations associated with rainfall land reflect the fact that it is held by a relatively large number of households—twenty-three, compared to fifty-seven for well land. Of these, twenty-one also use well land and generally own land of both types (correlation of .99 between amounts of rainfall land owned and amounts used, compared to a parallel correlation of .26 in 1965). In all likelihood, most of the rainfall land was retained by its owners and used on the expectation that it would soon be converted to well land, but the conversion had not yet been completed or had not been recorded.

CATTLE

Although the need for draft power was sharply reduced by the introduction of electric and diesel pumps, the village cattle population did not decline between 1965 and 1978. Increased fodder production is an automatic concomitant of increased grain production, and reducing the cattle herd would have let this gain go to waste. The better course was to increase the cattle population to a comfortable level while modifying the configuration of animals to increase their contribution to the level of welfare. There were more animals in 1978 than in 1965, but they were better fed and their configuration was oriented more toward providing food than power.

Table 3.5 gives the 1978 cattle census and compares the village working (adult) cattle populations for 1965 and 1978. The grand totals are not expressed in counts of individual animals but in the units the villagers use to estimate and compare both their feed and their work. The unit is variously called a "plow" (*hal*) or a "pair" (*jorī*). It represents what will be consumed as fodder or drawn as work by a pair of normal bullocks. The fodder allowance for a pair is fifty-four quintals a year, some dry and some green (for a fuller discussion, see Leaf 1972:46–54). Nonworking animals are fed less. A buffalo cow requires the same fodder as a pair of bullocks, as does a camel. A camel can by itself pull a Persian wheel or a plow (thus doing the work of a pair). The ox cow in the table is a female *Bos indica*, the same species of large white ox that provides the bullocks (adult, castrated male animals). An ox cow consumes half the fodder of a pair. A mule is also rated in consumption and work at half a pair, and a donkey at half a mule. A goat consumes one-fourth the allowance for a pair. Multiplying the consumption factors by the number of adult animals of each type and adding the products gives the result that the village in 1978 supported about 24 percent more cattle than in 1965. A comparison of this increase with the far larger increase in crop production (with the fodder it entails) is sufficient to explain the villagers' statements in 1978 that the previous shortage of fodder no longer existed.

The uses of cattle are related to their distribution among the socioeconomic groups of the village. Table 3.6 computes the changes between 1965 and 1978 in the numbers of cattle of each major type per person in the larger village groups.

In both periods, the buffalo was the major milk animal, giving almost

Table 3.5
Cattle Census, 1978

Cattle	*Brahmin*	*Jat*	*Mason*	*Gold-smith*	*Barber*	*Water-carrier*	*Potter*	*Cotton-ginner*	*Sadhu*	*Harijan*	*Sweeper*	*1965 total*	*1978 total*
Buffalo calf (f)	5	79	10	2	0	1	12	2	9	18	6	57	144
Buffalo calf (m)	0	8	0	0	0	0	2	0	2	2	0	6	14
Buffalo cow	7	129	10	2	2	0	16	3	8	38	5	147	220
Bullock	0	69	0	0	0	0	0	0	2	1	0	85	72
Camel	0	0	0	0	0	0	0	0	0	0	0	15	0
Cow calf	1	3	0	0	0	0	0	0	2	1	0	3	7
Cow calf (American)	0	0	0	0	0	0	0	0	0	1	0	0	1
Donkey	0	0	0	0	0	0	13	0	0	0	0	1	13
Goat (doe)	0	18	3	0	1	0	4	0	1	29	26	45	82
Kid	0	1	0	0	0	0	0	0	0	0	0	15	1
Mule	0	0	0	0	0	0	4	0	1	0	0	24	5
Ox calf	1	6	0	0	0	0	0	0	2	0	0	13	9
Ox cow	1	7	0	0	0	0	0	0	0	5	1	17	14
Pig	1	0	0	2	0	0	0	0	0	0	13	13	16
Ram	0	7	1	0	1	0	1	0	1	6	0	3	17
										Total adult "pairs"		234.25	290.25

Table 3.6

Ratios of Major Cattle per Capita, 1965 and 1978

Cattle	Brahmin	Jat	Mason	Gold-smith	Barber	Water-carrier	Potter	Cotton-ginner	Harijan	Sweeper	Sadhu
Buffalo cow											
1965	0.800	0.250	0.160	—	0.250	—	0.110	—	0.140	0.020	0.400
1978	0.368	0.346	0.167	0.154	0.250	1.000	0.167	0.250	0.123	0.074	0.500
Bullock											
1965	—	0.208	—	—	—	—	—	—	—	—	0.600
1978	—	0.185	—	—	—	—	—	—	0.003	—	0.125
Cow											
1965	—	0.020	0.020	—	—	—	0.050	—	0.080	0.020	—
1978	0.053	0.019	—	—	—	—	—	—	0.016	0.015	—

twice as much milk per year (over a longer lactating period) as the cow. Buffaloes are owned by all groups, especially by families with children.

In 1965 a good milk yield was said to be about 6.5 kilos a day, with a nine-month period of lactation. In 1978 the milk yield was stated as being 10 kilos at peak, down to 1 kilo at the end of the year. The increase of almost 50 percent in the total number of milk-producing cattle therefore represents a more-than-proportional increase in total milk production.

The increased milk yield per animal is due partly to the increase in fodder itself, but mainly to the better quality of the animals resulting from a government artificial insemination program that began a few years before my first study. This program in turn partly explains the absence of male buffaloes in the village. They are not generally good for work in this area (it is too hot and dry for them) and there would be no profitable way for anyone to keep one for stud service. In 1965 an unowned buffalo bull did come wandering around occasionally. He was allowed to graze where he was wanted for stud and driven away where he was not. In 1978 when I asked several farmers where their buffaloes came from, the answer was always regional cattle fairs.

Female goats are used for milk and kids, and then for meat. The efficiency of goats as milk producers is less than that of buffaloes, but it is very good. Male goats are used only for meat. Hence the goat population as a whole represents the equivalent of a further ten buffaloes to the village milk supply and more than a doubling of the total number of animals available for meat (neither buffaloes nor cows are eaten). The increase in the number of rams is probably the most dramatic indication of the increase in general welfare. They are the preferred meat animal, but in 1965 the shortage of fodder made it impossible to feed an animal for two or three years just for slaughter for an ordinary meal or for market. (No one will kill a kid.) Thus young males were ordinarily sold out of the village. In 1978, with the fundamental need for power taken care of in other ways and with an adequate milk supply in relation to local dietary needs, there was no reason not to raise them.

Bullocks and camels are draft animals, owned only by families engaged in farming. The decline in their numbers represents a total reduction of plow units by 39 percent, a direct consequence of the adoption of mechanical pump engines.

Cows are kept by very poor people as inferior milk animals (because they can sometimes be obtained free) and by farmers to produce new bullocks.

Mules and donkeys are all owned by Potters and are used for carrying their heavy wares in compact packs. A mule can carry about two quintals, a donkey one.

In general the animals are still kept only where they are needed. No one, for example, retains cow calves only because they are "sacred." Yet the uses for animals in 1978 were on the whole less pressing than previously and were more often related to comfort or long-term need than to short-term need or hard necessity.

HUMAN POPULATION

One of the often-asked questions concerning the Green Revolution is whether the labor requirements of the new crops necessitate such large population increases that the benefits are offset and no real gain per capita results.

Determining the number of people in a village is not as simple as it might sound. People are not eggs, and villages are not crates. My 1978 census of the village discovered three families that my 1965 census had missed and showed the need for more precise criteria for saying who was in the village and who was not. Applying the new criteria to both censuses consistently required some reclassification of the 1965 data, so the figures presented here will not precisely match those in *Information and Behavior in a Sikh Village*.

The basic problem is that people have different degrees of associations with different places. Some live and work in the village and seldom travel out. Others travel out daily to work or school, but return for meals and to sleep. Others spend months or even years outside the village, but have no intention of leaving permanently and remain outside only in order to provide for families who stay behind. Still others, notably girls who marry out, have left the village in theory but in fact often come back and in principle always can come back "temporarily." Finally, there were a few people in 1978 who commuted to the village to work on a daily basis but had their primary residences elsewhere.

A three-way distinction between resident villagers, nonresident villagers, and nonvillagers produces the fewest ambiguous cases. Residents are those reported as belonging to village families and as taking meals regularly in their village houses. Nonresident villagers are those who are reported as belonging to village families, but not taking meals with them regularly at the time of the census and usually also as having a definite place of temporary residence other than the village. Finally, nonvillagers are those reported in the census but not considered part of village households. This category would include both out-married daughters and those who commute to the village for work.

In 1965 there were 871 villagers, resident and nonresident. There were 812 residents. In 1978 the number of villagers had risen to 1,076, while the number of residents had risen to 975. The increase of 20 percent in residents is obviously not close to the increase in food production. Nor is the increase of about 23 percent in total villagers reported, even though this figure is undoubtedly inflated by the 1978 procedure of specifically asking about each person listed in my 1965 forms.

The population increase was not uniform across all occupational groups. This fact reflects changes in the division of labor of the village and will be described in the next chapter.

HUMAN DIET

Although by 1978 the villagers were growing proportionally more of their crops in response to external demand, almost all of what was eaten in the village was still grown there. Very little was brought from outside. Thus human diet still set lower limits on the animal and plant populations required in the village ecosystem.

One set of reasons for maintaining an internal subsistence base in the village is economic, based on the great difference between costs of local production and retail prices in the area. This will be made clear in Chapter 5.

The second set of reasons is ecological. The village diet is in fact quite adequate nutritionally, and for most villagers it is also the one they prefer. The village recipes do a remarkable job of building a well-rounded diet that is at once cheap, easy to prepare, and tasty enough to stay with day after day, year after year.

Management of the household diet is the responsibility of the senior woman. She has to plan to have the necessary raw materials on hand in the correct proportions each season and each day, and she has to oversee the rates of consumption and make sure that what is set aside for local use is coordinated with what is sold in the market. This is a complex task and it helps explain why the diet contains relatively few distinct dishes. The main types of preparations are *roTī, sāg, subjī, dāl,* meat, and sweets.

RoTī usually consists of flat, unleavened, soft wheat or cornmeal cakes cooked on a hot griddle without fat. *Sāg* is a cooked green, with oil, onions, garlic, and spices. *Dāl* means beans, usually lentils or chickpeas, cooked with onions, garlic, water, and spices in a thick soup. *Subjī* means vegetables (but not *sāg*) and usually consists of cauliflower, potatoes, peas (alone or in combination), or squash, cooked in the same general manner as *dāl*. Meat, too, is cooked like *dāl*. Sweets take many forms, but sugar is basic to all. There are now many good recipe books available in English that can provide more detail.

A meal normally consists of *roTī* and one or two side dishes, such as *sāg* and *dāl* or *subjī* and *dāl* or meat, along with tea and some sweets.

The tea is made with substantial amounts of milk and sugar. In a day, a person would normally eat a light snack for breakfast (*roTī*, sweet, and tea), a meal for lunch, another light snack ("tea") in the afternoon, and another meal in the evening.

For the 1965 base period, I calculated that the nutritional contents of the stated dietary allowances for a working man would have provided approximately 3,141 calories a day. This would include 81 grams of protein, 85 grams of fat, and 562 grams of carbohydrates (on the basis of USDA analyses of the recipe ingredients in the form found in the United States). Minerals and vitamins included 1,390 milligrams of calcium; 3,085 milligrams of phosphorus; 30.7 milligrams of iron; 1,400 milligrams of potassium; and 22,500 international units of Vitamin A. Allowances for women and children would be scaled down from the allowance for men, and most people who were not working strenuously in fact probably ate less than the allowances.

In 1978, with the additional milk, wider range of vegetables, meat, and occasional eggs, the diet was probably a bit better. Certainly it was better for children, and the villagers unquestionably enjoyed the additional variety. The result was that in 1978 as in 1965, there was no gain to be had in dietary quality, as such, by importing foodstuffs from elsewhere.

HOUSING

To urbanized Westerners it is difficult to convey the importance of housing in the pattern of village life. Economists conventionally consider housing a form of consumption, not capital (even though noneconomists recognize that houses are often important investments). This suggests that a house is something obtained out of one's income, rather than a basis of income. In the village, houses are capital preeminently. They are places to work, to study, to store food and goods, and to rest. They are the basis of the village system of resource management, which will be described in the next chapter, and through it are also irreplaceable parts of the village ecology. The human population in 1978 would have been worse off, not better, if housing was the same as it had been in 1965.

Between 1965 and 1978 there was a substantial improvement in the total stock of village buildings. Figure 3.2 is a map of the *abādī* showing the areas of housing in 1965 and 1978 along with a rough indication of where members of each socioeconomic group live. The areas indicated as housing are between walls but not necessarily under roofs.

Forty-three percent of the households censused in 1978 reported remodeling or new construction of at least one housing unit since 1965. It is diffi-

Figure 3.2 Map of *Abādī*, 1978

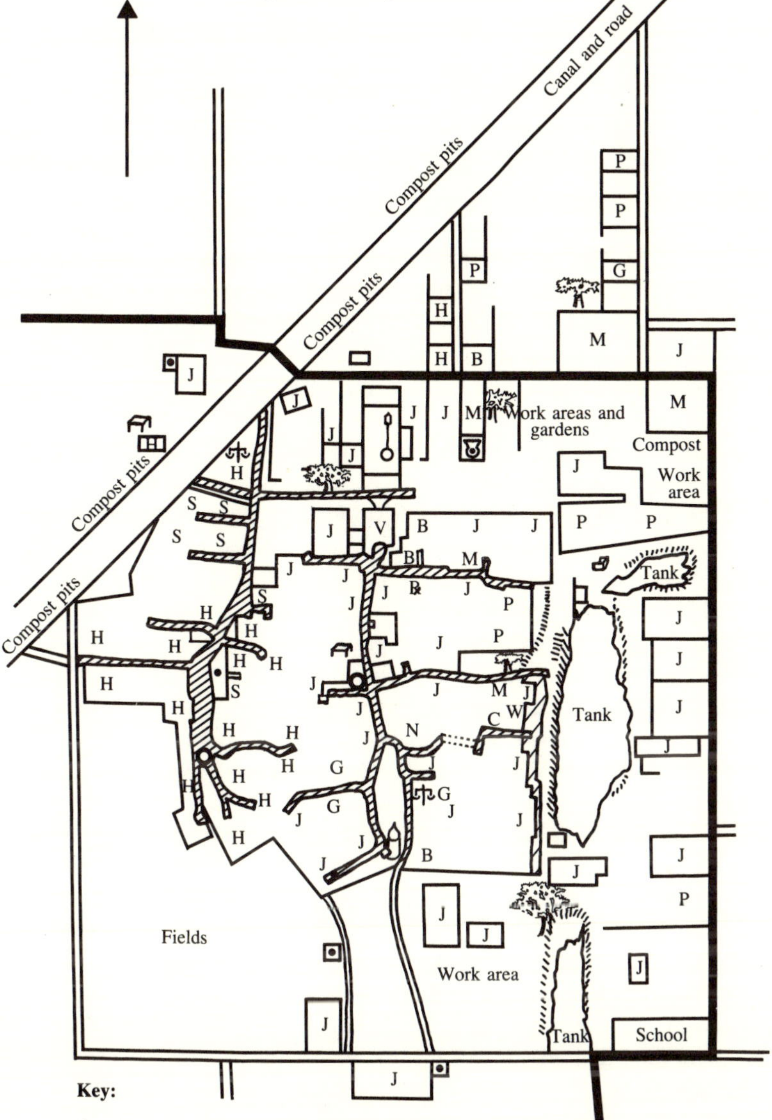

Key:

(symbol)	Sikh temple	(symbol)	Hindu temple		
B	Brahmin	(symbol)	mill	•	private well
J	Jat	(symbol)	shop	V	village gate
M	Mason	℞	doctor's office	W	Water-carrier
G	Goldsmith	(symbol)	dharmsala	C	Cotton-ginner
P	Potter	(symbol)	pottery kiln	H	Harijan
N	Barber	o	public well	S	Sweeper

cult to estimate the sizes of individual village houses accurately because of their irregular shapes and because they share many common walls, but by drawing my *abādī* maps for the two periods to the same scale and comparing the total areas occupied by housing, allowing for changes in architectural style, it appears that the total area under roofs has increased by about 20 percent.

Within the houses, there has been an important addition to the techniques for grain storage. One of the main sources of India's persistent difficulties in assuring a regular internal food supply is that it has no system of large-scale grain storage and movement comparable to the system of interrelated commercial silos and railheads in the United States. There is no way farmers can place grain in commercial bulk storage for later sale. They either keep it or sell it. If they sell it, they do so in sacks, not in bulk, and the dealer who buys it often has only a warehouse, not a silo, so no real preservation is possible. If farmers keep the grain, they have to do so in their barns and storerooms. As a result, loss rates have historically been rather high, although older villages, like Shahidpur, present fewer problems because the *abādī* gradually rises on a mound above the natural plain and hence is away from pests and the most frequent floods.

In 1965 the best village method for grain storage was to build a concrete floor covering the entire bottom of a brick-walled room with a tight door. Wooden pallets were placed on the floor and sacked grain was stacked on them. Beginning about 1972, large galvanized steel drums became available commercially for about Rs. 325 each. They are cylindrical, about two meters high, with a removable top and an outlet at the bottom from which a plate can be removed to allow grain to spill out. Each holds about ten quintals (one metric tonne) and can be sealed airtight and fumigated. Two years later it became possible for farmers to apply for a 50 percent government subsidy on the purchase price. The result has been a great improvement in storage capacity and almost complete freedom from the fear of natural pests and possible weather damage that formerly forced farmers to choose between selling a full harvest early at a relatively low price or holding for a higher price later at some risk of loss. Farmers reported that all landowning families had the metal drums in their houses. Each house I visited personally had at least one, usually two.

Housing for cattle has been improved almost as much as storage for grains. In 1965 cattle (especially buffaloes) were often housed indoors, but were almost always kept over bare earth. In 1978 almost all buffaloes were kept on brick floors, both to improve their own health and to assure cleanliness in the milk. The impetus appears to have come from a government milk scheme, which will be described in Chapter 5.

The housing stock of 1978 was, in a literal sense, a material embodiment of the greater intensiveness of crop production and the corresponding greater density of human and cattle populations, in a wider institutional setting.

TOOLS

The village tools are the means by which people control the natural processes that tie the various crops and animal populations into a single ecological system. The best way to describe the way they have changed is by the major tasks they are used to accomplish. These are water lifting, soil manipulation, transportation, threshing and milling, fodder preparation, and food preparation.

Water lifting, as noted, has been one of the areas of most dramatic change. All of the forty-eight mechanized wells in the village used centrifugal pumpsets, although other types were available in the region. The wells themselves were of two types: tube wells and ordinary shaft wells. Tube wells consist basically of a pipe about six to eight inches in diameter (15–18 cm), driven down ten to fifteen meters below the ground water level. The pipe ends either in a long, perforated strainer or an underground tank, depending on the soil. Within the main pipe is a smaller one, which is attached to the pump proper. The large pipe is open at the top to allow air to enter the chamber that collects water; water is drawn up the inner pipe.

The shaft well is always one that previously had a Persian wheel: masonry walled and about three meters in diameter. The newest well shafts go down to about twelve meters; the older ones are much deeper. A pump is attached to a shaft well by mounting it at the ground surface or in a pit next to the brick wall and dropping the intake pipe into the shaft, as far down as possible, with a small strainer (much simpler than that of a tube well) on the end.

Each type of well has advantages and disadvantages. The tube well with a cavity has a high rate of flow, a reasonable cost, and lasts indefinitely. It can often be pumped twenty-four hours a day. But it can only be used in firm soil where the underground tank can be inserted. The tube well with filter has a lower rate of flow, costs more, and can be put anywhere, but in this area it lasts only for seven years, until the strainer sleeve becomes blocked or corroded (in areas with different soil chemistry its useful life could differ). Although the shallower shaft wells run dry in a day of heavy pumping during a dry season, they usually refill by percolation overnight. They are already in place, last indefinitely (some shafts in the village are perhaps three hundred years old), and the only installation expense required is that for ten or fifteen meters of pipe. The new wells are all of the tube type, but in most cases where farmers had shaft wells with Persian wheels they have mounted pumps on them.

To run the pumps, there are three types of motors, distinguished by their fuels: electric (called *moTaR*), diesel and "black oil" (both called *enjin* or *dījal*)—also a type of diesel. All are used in the 5 HP size. Light diesel is by far the most popular. In 1978 there were about four electric motors mounted on wells in the village. Subsequently Surindar wrote that many more had been installed.

Once a motor is installed, water lifting is only one of its uses. Ingenious systems of pulleys and belts link it to a wide range of other implements, some formerly run by human or draft power and some entirely new. The older types of tools include cane crushers, centrifuges for separating crystal sugar from molasses, and fodder-choppers. The most significant new machine is the general-purpose chopping and threshing machine called a drummy.

In the census, a total of forty-four households reported owning a total of fifty-one motors. Four of these families (those of the Mason, the miller, a farmer, and a Cotton-ginner) had engines not used with well pumps. Since two of the fifty-two wells had been reported as having Persian wheels and one that I saw had neither Persian wheel nor motor, this leaves the motive power on two wells unclear. Since I did not collate this information while in Panjab, the discrepancy was not checked. Most likely it reflects uncertainty over how to report broken motors or parts of motors not working that could be assembled and made operational if needed.

Soil preparation included plowing, sowing, leveling, and considerable hoeing, weeding, and digging. All of the equipment used for these purposes in 1965 was still in use in 1978 but there was much more of it, especially of the smaller hand tools. In addition, however, there were four new tractors.

The type of plow used in 1965 is shown in the illustration section in the photograph of Gajjan Singh (who was to be Secretary of the Cooperative Society in 1978) breaking ground for winter wheat. This plow was not the ancient type but rather an improved version of it, made commercially. As can be seen, it used a steel blade that turned up a row of earth on either side of the furrow being cut. The usual depth of the furrow was about 18 cm, but this could be adjusted by changing the angle between the lower beam holding the blade and the drawbar attached to the yoke of the cattle. The plow was easily steered by tilting the almost-vertical handle. A drill for seed or fertilizer was a simple tube with a funnel at the top, attached along the back of the handle. The use of such a drill required two men: one to steer the plow, one to drop in the material at the proper rate. If a farmer wanted to apply fertilizer and seed at the same time, it was necessary to have three people. One would drive the team and steer the plow. The second would drop fertilizer into the drill, which would be set to carry it to the bottom of the cut. The third would drop the seed by hand just behind the plow so that it would fall into the earth flowing around the blade and falling back into the furrow on top of the fertilizer. If done just right, this left a row of fertilizer at the very bottom of the furrow and a row of seed about 4 cm above it, but still in the furrow and lightly covered with soil. The next pass would turn half the soil from the first furrow back in on top of the seed and fertilizer, and add about the same amount of soil from the new furrow above it, thus neatly burying the seed.

The main equipment that accompanied a plow was a flat rake and a level used to break up clods and smooth a plowed field before sowing. The rake and drag were often operated by one person riding on it and steering the cattle and several others standing alongside for added weight.

The plow could easily be carried by one man, although more often it was taken to the field by the cattle themselves, hung inverted over their yoke. The yoke itself was, as can be seen, a single wooden bow shaped to stay on humped cattle by the force of gravity alone; no harness or hitches were used. The cattle were steered by a light string attached to a nose ring; they would move forward when prodded (not beaten) with a small stick in their more tender regions near the tail. The plow was attached to the yoke by a simple loop of rope, as can be seen. When cattle or mules were used in American farming, it was considered a good half-hour's work to hitch a team to a plow. In Panjab, the same job takes about three minutes.

For all of its elegance, by 1978 this type of plow was recognized as having at least one major drawback: it was too slow. The need to let the land "rest" between grain crops, the increasing use of intercrops and crops with longer growing periods, and even the reduced number of draft animals, all put an increasing premium on plowing speed. This is where tractors became most important.

In 1965 the cost of tractors was considered prodigious. In 1978 there was one of 20 HP, two of about 50, and one somewhere in between. A 50 HP machine reportedly could plow the same area in a half-hour that a bullock team could plow in a day, and it required only one man for all operations.

The plow pulled by the tractor was a local three-bladed type called a *triphālī*. Other high-priority attachments included large circular iron frames (looking something like small Ferris wheels) that bolted to the outside of the rear drive wheels to increase the area of disturbance for puddling rice fields, and two- or four-wheeled utility trailers. Small combines were also available and village farmers occasionally hired them, but no one owned one. Netar said they made the straw useless as fodder.

Tractors were also used occasionally to operate well pumps, and with their trailers were commonly used on the roads both for carrying produce and for carrying groups of people, such as wedding parties.

In 1965 transportation was accomplished by three major means in addition to walking. There were about 120 bicycles (one per house) and about twenty-five high-wheeled oxcarts. For heavy loads or large groups, trucks or buses would be hired from towns, provided the dirt roads to the village were not flooded. In 1978 the number of bicycles had increased to 210. The traditional oxcarts had declined in number to eleven, but there were twenty six of a new type that had just been coming into use in the region at the time of the first study (Leaf 1972:258). This type was made on a truck differential, with rubber tires, and could carry twice the weight of the older cart with the same draft power.

In addition to the bicycles, oxcarts, and the four tractors and trailers, villagers owned four motorcycles and one automobile. One family also owned a ten-ton truck, but this was not kept locally. If one considers bicycles and oxcarts alone, it is obvious that the village transportation capacity in 1978 was over twice what it had been in 1965. When the motorized vehicles are considered as well, a threefold increase is probable.

Threshing is not the same for all crops, and cannot really be described in general terms, but in 1965 it was almost always done either by hand or with bullocks and a drag (for wheat). Maize was allowed to dry on the plant, shucked, and stored on the cobs. The village Mason, Gian Singh, who owned a portable (but not by any means light) diesel engine, had just assembled a system for power threshing of wheat, which he had hired out in the summer of 1964 with some disappointment. (The story was that only a few farmers were interested, and one of them had cheated him out of a day's fee by using the machine to chop up the ropes that would have served as evidence of the number of bundles threshed.) He was hiring it out again in the spring harvest season of 1965 to a few people he trusted more completely, and was waiting to see if it would prove worthwhile. No one at the time appeared certain of the future of such technology.

For milling wheat, the village miller used a large horizontal stone mill powered by a diesel engine. For removing maize from the cobs and cracking it to usable size, Gian Singh (the same who assembled the thresher) had another device powered by the same engine. In 1978 the miller's equipment was basically the same, but for threshing wheat, barley, millets, and *dāls*, farmers generally used drummies.

Fodder preparation consists of chopping and mixing. Chopping improves digestibility of materials otherwise too coarse. Mixing allows the combination of many materials that lack some of the qualities of a good fodder. For example, high-bulk materials low in nutrients (like wheat straw and rice straw) are combined with low-bulk materials high in nutrients (the husks and stems of lentils and peas, for example, or millet grain).

In 1965 fodder chopping was mainly done with hand-powered wheels. The first improvement was to modify the hand chopper so it could be driven by a belt attached to a second small, geared wheel. The second improvement was the aforementioned drummy. In 1965 farmers estimated that there were a total of eighty chopping wheels in the village. No one mentioned drummies. In 1978 there were more wheels (they were so common there was no interest in listing them on the census). Five drummies were reported, including one with an engine permanently mounted on an oxcart, evidently for hire. More probably existed. In 1965 it was usually necessary for at least two people, and usually three, to spend an hour or two every day bringing material to the chopper, cutting it (one turning the wheel while another fed in the material in small quantities), and mixing it. The drummy, or power wheel, could be fed by one person and do the same job in perhaps a fifth of the time.

Of all the major activities, food preparation has changed the least. Food for a family is still cooked on a single small fireplace (*chūlā*) in heavy brass pots and involves a great deal of personal attention and manipulation. The main change concerns fuel.

In 1965 the principal cooking fuel for all villagers was dried cow dung cakes, supplemented by corncobs in the winter if the family grew maize. For fires that did not need to burn for a long time, or for intense bursts of heat, grassy weeds or crop parts such as roots or cotton stalks were used. This was also a necessary fuel for people without land and/or enough cattle to supply the necessary amount of dung. Cooks spent a great deal of time blowing through lengths of pipe to start or intensify their fires, and on calm days the smoke often hung over them in acrid clouds. Kerosene pressure stoves, charcoal, and coal had all been mentioned by villagers at one time or another and were considered very convenient but too expensive to use in the village. No one mentioned wood.

In 1978 one more fuel was known but not used—biogas. Biogas is methane, created by allowing organic waste to decompose in a large tank sealed so that air can not reach it. Maintaining such a tank requires the collection of large amounts of waste that would normally be composted (proponents of such projects insist that its value as compost is not thereby lost). The gas created is carried to separate village households in pipes and used in small gas stoves. There were, according to Mohan Singh, four demonstration plants within eight kilometers of Shahidpur. Some of the villagers had gone to see them but were not particularly impressed. One reason was the unpleasant smell and another was the amount of land required for a waste sump, but the more serious problem was that while this process gave a great deal of gas in the summer, when it wasn't much needed, in the winter it would provide enough gas for small families but not enough for large ones. This would have meant that the old *chūlā* could not really be abandoned, and if the fuel collection for the gas plant was truly thorough, it would also have meant that the villagers would be faced with an absolute fuel shortage.

But the lack of interest in biogas did not mean there was no interest in change. In 1978, much to my surprise, perhaps a third of the families were regularly cooking with wood, even though it cost four times what it had in 1965. Dried cow dung was still the main fuel of poorer people, but there was much more of it. I saw no stacks of stems and roots before the fireplaces that formerly had them. This may seem insignificant, but when one considers the hours spent daily in every house gathering fuel and preparing food, it represents a very substantial increase in convenience, improvement in working conditions, and saving in time.

THE ECOLOGY AS A MESSAGE SOURCE

After all this it may seem overly bookish to return to the theme that the ecology is a message source of high information potential, but this is directly related to three points of great importance: the magnitude of the changes has been great, the changes have not caused any loss of stability in the system, and they have been brought about by many people acting more or less independently.

The loose organization (hence high information potential) of the ideas of the ecology may be more obvious to a writer of an account such as this than to most readers. The writer is necessarily more aware of the alternative ways the account might have been organized and the alternative information that it might have included. I put land first and proceeded generally from the things that were larger and took longer to change to those that were smaller and changed more easily. But most villagers who learn the system start with a few simple operations and slowly learn more and more, with a general sense or overview of the system coming much later. There are strict relationships, of course, but there are so many of them that no one has a determinate effect on the organization of the ecology as a whole. This is why each person has a great deal of freedom to choose the aspects of the system he will work with, and what he will do with them—and this in turn is precisely what allows the system to respond rapidly to external opportunities.

There was no change in the basic biological processes that underlay the village ecology between 1965 and 1978, but in 1978 there was a definite sense that the range of ways to exploit these laws had become wider. There was more freedom in the system; its information potential, in a formal sense, had increased. More was produced; there were more real resources and more ways to seek and dispose of them without risk of disaster.

4

CHAPTER

The Division of Labor

The system of management, or division of labor, is the system of relations by which goods are bound to individuals and individuals are bound to each other in distinct and autonomous decision-making units. It is a system of many fewer elements than the ecology: about thirty basic concepts and about one hundred and fifty actual decision-making groups that each villager must know. Yet it is a system of great flexibility.

The standard Panjabi term for management (the noun) is *parband*; a system of management units is *bandobast*; and a manager is *parbandhak*. The root *band* means "tied" or "bound" and the imagery exactly conveys the idea. In addition to the general concept of management, there are concepts for the specific management groups that will be described. Other important general concepts include organization, strength, property, possession, rights, rightsholders, cooperation, competition, unity, decision, coordination, opposition, obligation, purpose, and goal. The definition of each of these involves one or more of the others. Some of these concepts are defined in law and sanctioned by courts. These include the ideas of right, rightsholder, *bandobast* (in the name of the legal record of land titles), property, possession, and all three types of rightsholding groups: the cooperative, the panchayat, and the family group (*parivār*). Other concepts are sanctioned less formally (but no less forcefully), while some are sanctioned both ways. All these rather narrowly managerial terms are used along with a very large number of additional concepts that are primarily defined in other contexts, such as kinship and party, and are recognized as being used here somewhat loosely.

There is no particular necessity that the legal sanctions and the informal sanctions coincide. Indeed they often do not: ideas backed by law and the courts have consistently been at variance with those supported by local sanctioning arrangements such as the force of individuals, the pressures on families, or the pressure that can be brought to bear by families organized through larger kinship groups or through parties.

Some of the legal measures that have been taken up in the village system will be described here. Other ideas which legal authorities attempted to impose but were resisted by villagers will be described in the chapter on kinship, for reasons that will be given there. These attempts, even though unsuccessful, were a source of worry in the village and region between 1965 and 1978 and have been a major cause of the increase in political unrest in the state since then.

Within the almost-constant conceptual framework, most of the actual units in the village division of labor had changed at least somewhat in structure and in the resources they controlled since 1965. The pattern of mutual relations between them changed correspondingly.

THE VILLAGE PANCHAYAT

The term "panchayat" is most literally translated as "council of five." It is commonly said that there were, in the past, panchayats of many sorts: in the Sikh religious community, in villages, and in castes. Such panchayats, where they can be documented at all, seem to have been quite ad hoc and informal. Standing village panchayats established by law, like the one in Shahidpur that is of interest here, are of recent origin.

In Panjab some village panchayats have been regulated by law since 1921, but they were not established universally until the Panjab Gram Panchayat Act of 1952. (*Gram* is a Sanskritic term for "village.") Shahidpur's panchayat dates from this act, which provided for panchayats in single villages where the population was over 500 and for panchayats of combined adjacent villages where the populations of each or some were less, with a minimum of 500 per panchayat. Each panchayat would have between five and nine members. Election was by universal adult suffrage. "Scheduled" (listed) castes, considered weaker than the others in that area, had seats reserved for their members in proportion to their population. Only members of such castes could vote for the candidates for these seats, although nothing prevented a person of scheduled caste from running for an open seat or voting for candidates for open seats. The law also provided that at least one woman would always be seated. If none was elected to an open seat, the woman who received the largest number of ballots from among those running was to be coopted. If none ran, the panchayat itself could coopt a woman from the area who was willing to serve. One member of the panchayat (panch) was directly elected as sarpanch (literally "head panch"), who chaired the meetings and kept the records. Meetings had to be public and held at least monthly. Decisions were by majority vote; in an equal division, the sarpanch had an additional, or casting, vote.

The 1952 act has been updated by a long series of amendments, which have generally increased the panchayat's powers (by small increments) and strengthened the hierarchy of indirectly elected bodies (panchayat samitis) above the village panchayats, paralleling the administrative structure of the state, that is referred to as panchayat raj.

By 1965 panchayats had considerable legal power. In practice, however, it was largely unused—and this itself was an important feature of the village management system. They sat as courts of initial resort for all three main branches of the Indian judicial system: revenue, civil, and criminal. They could try a long list of minor crimes, such as assault, public drunkenness, and theft under Rs. 250. They could hear civil suits concerning similar small amounts. They had the power to collect certain types of local taxes, mainly "hearth taxes" (a tax per household). They could regulate the distance from the *abādī* at which certain activities or trades might be practiced, like hide tanning. They could regulate fairs and periodic markets. They could set aside land and seek grants for public schools and clinics. They had certain supervisory powers over local revenue clerks (*patwārīs*). They could hear charges of fraud involving weights and measures in local businesses, and they could levy fines of varying but substantial amounts and impose jail terms for up to three years. The panchayats were also the owners of record of village public lands, including cremation grounds, roads, and school sites. Panchayat procedures were less formal than those of higher courts. No one was permitted to bring a lawyer to plead for them, but they were courts of record.

Finally, members of panchayats were official witnesses, together with *lambardārs* (hereditary tax collectors for groups of related landowning families in the village). They were legally obligated to report certain classes of major crimes (murder, rape, and robbery), they were legally empowered to act as witnesses in both civil and criminal cases of all sorts, and they were compensated by the state for time spent doing so. In addition, they were generally called upon to serve as witnesses in ordinary major transactions, such as the sale of land, where they would attest to the identities of those appearing and to the accuracy of the deed description. In 1965 this power of the individual members of the panchayat clearly overrode all the powers that the panchayat had collectively. The major change by 1978 was that some collective functions had also come to be prominent.

In conversations I had with regional scholars and officials before beginning work in the village, the panchayat had often been connected to parties and to the idea of corruption. Accordingly, I had mentally marked it as a probably sensitive topic, one to be pursued only if villagers themselves brought it up in the normal course of our discussions.

As it happened, I was immediately introduced to individuals who were explicitly described as panches, the sarpanch, and former panches, and they did indeed prove to be quite prominent in village affairs. Yet in a specific and quite straightforward sense, during my entire first study I never encountered a decision of the panchayat as an autonomous body. An example will explain what I mean.

One afternoon late in October, a few weeks after I had settled in the village, I was in my room in the upper story of the gurudwara trying to memorize some Panjabi. Rather suddenly, my study was interrupted by loud, angry voices from the courtyard below. I waited but the argument did not subside. Nor did anyone come up. Finally there seemed no choice but to go out.

From the veranda I saw that just in front of the building, on one side of the central walkway, there was a large cot. Several men were sitting on it and a number of others were standing in front of them in a semicircle. Some of these were participants in the argument; others seemed to be bystanders. As I came down the steps and was noticed, the argument quieted. Several of those on the cot made a point of calling me over in the usual hearty and friendly way and bid me to sit with them, on one end. I did so, quite relieved, but the conversation did not resume. So I asked what was happening. In reply I was shown a watch by one of the men on the cot and asked its value. It was Indian-made, of a type that sold for about Rs. 75 when new. I guessed it would be worth about Rs. 50. Immediately several of the men on the cot and several others resumed shouting at one another in language I could not follow. Two almost came to blows, with the participant who was standing being restrained by others. Minutes later, with the argument still heated, those on the cot abruptly stood up, picked up some records that had been spread out there, and walked away. Most of the rest dispersed as well.

Naturally I was curious to know how my response could have had this effect, so I asked a couple of the bystanders I already knew. The meeting had been "the sarpanch's court" and the issue was an accusation of theft that had been brought up after a previous decision. Those on the cot and the man who had had to be restrained in the argument were the panchayat. The watch was the object of the charge. It had been in the possession of one villager and later appeared in the possession of another without benefit of sale. When its whereabouts became known to the original owner, he brought the charge. None of these points were disputed and none were debated. The debate had been about jurisdiction. The majority members of the panchayat held that the council could only decide accusations of theft if the amount involved were less than Rs. 50. (This had been the limit before 1960, but was not the limit in 1965.) The only question they would consider was whether the watch was over or under the limit. By chance I had named the pivotal value. The majority then construed my statement as favoring their position, namely that the case was out of their jurisdiction. They argued that the complainant should take the matter to the district magistrate—a remedy much too costly to be justified by the value of the watch. The minority members disagreed with the jurisdictional argument and held that it was also important to consider the questions of right and wrong in the case.

All of this may sound like a decision, but when I asked about its seeming arbitrariness, it was immediately explained by the same bystanders that those involved acted as they did because of the party alliances of the village. The majority in the case were all members of the dominant village party; the remaining two panches were members of the minority party. The original owner of the watch supported the minority party; the thief supported the majority party. The members of the majority party in the panchayat invariably supported those in their party; the minority party necessarily supported those on their side.

Although I was inclined at first to take this as a stereotyped statement that would be forgotten in actual ongoing behavior, the pattern always held in subsequent actions. Perhaps this was what urban critics would have called corruption. From the village point of view, however, such a characterization would have seriously missed the mark.

The key to the behavior of the panchayat lies in the power its members have to influence the transfer of land. This power is not part of its legal jurisdiction, but flows out of the role of the panchayat members as official witnesses. Because of the extreme importance of land in the village management system, if a person were deprived of it by the action or inaction of a member of the panchayat, he would be most unlikely to shrug and accept some formula like "he was only doing his duty," or "he was only following the law." To deprive a person of land is not only to injure that person but to injure or ruin his family as well, and it is difficult for a person so injured, who sees himself as having done no wrong, to be detached about such treatment. Because of this, members of the panchayat were in the end virtually forced to act as they were thought to act—to exercise their power as witnesses strictly as an aspect of the village party alliances. A person in a dispute over land who was of one party would be sure that the panchayat members and other official witnesses of his party were present in court to provide testimony favoring his case. His opponent would be equally sure to have authoritative witnesses and others from his own party. Whoever had the most witnesses was expected to win (assuming the stories were equally consistent and plausible). It was impossible, by the logic of the concept of parties (not the law of panchayats), for people on both sides of a dispute to be of the same party. Disputes like that over the watch, where the stakes were trivial in comparison to a dispute over land, were necessarily seen as tests, or manifestations, of the loyalties that were important in the larger quarrels. There was no logical way for a panchayat member to separate his legal power as a witness from his other legal powers, so the party ideas were established as the ideas the panchayat member used to order his activities, while the legal ideas were simply not imputed as basic.

I have no way of knowing whether conflicts over land were especially intense during the time of my first stay, but villagers had two cases before the Panjab High Court and several score before lower courts. Most of the lower court cases were related to a procedure called consolidation of holdings which had been going on in the village as a result of the new canal minor being laid in 1962. One High Court case concerned title to land held in the name of the village dharmsala (rest house). The other, which attracted more attention, was a petition by twenty-one village landowners to quash the consolidation proceedings. Criminal charges were sometimes also laid as tactical adjuncts to direct disputes over land. A case before the District Court at the time of the incident over the watch concerned the Block Development Officer (the same who had introduced me to the area). He had charged the village sarpanch and one panch, of the majority party, with assault in order, it was said, to have them jailed and thereby removed as witnesses in other cases under adjudication. A countercharge of false arrest was then filed by the two villagers to threaten the same effect in reverse. (Evidently this was why the BDO did not bring me to the village personally.)

The practice of land consolidation began in Panjab in the 1930s. The law current in 1965 dated from 1948. The legislation was based on the belief that traditional holdings, through inheritance over time, became divided into many scattered tiny portions that involved much unnecessary movement between fields, wasted much space in field boundaries, and blocked mechanization. Its aim was to group such fragments into one or two major blocks, and thereby promote efficiency and create more land and better returns for all. In fact, the land of Shahidpur was not particularly fragmented and field boundaries took almost no space at all, but that is another story.

The consolidation procedure begins with an evaluation of the land of each farmer in terms of "standard acres." A standard acre is an acre of average productivity. An acre of land that is half again as productive as the average would be evaluated at 1.5 standard acres, and one half as productive would be 0.5 standard acres. Next, all the lands are pooled (on paper) and the amounts necessary for public purposes, such as canals, are taken out. The remainder is redivided so that each person gets a share equal to his *share* of the original total amount. Thus, for example, if a person had 10 standard acres out of a village total of 400, and if 40 acres were taken out for public purposes, he would get back only 9 standard acres. Under the rules this should be located, if possible, in one block in the area where he previously held his largest fragment. Everyone agreed to the general fairness of the idea. The problems arose where evaluative judgments were called for: in assigning the original worth of the land and in deciding where the new area should be given.

The procedure is conducted by a relatively high-ranking and generally unimpeachable Indian Administrative Service officer, but he does not have the necessary intimate knowledge of the land of each village. Accordingly he is advised, formally, by a local committee. Shahidpur's committee had thirteen members, including all of the panchayat.

The panchayat serving during my first study had been elected in 1964. The next election was held in 1972 and the group returned then was still serving in 1978. The 1972 election separated Shahidpur from a neighboring village and also, for the first time, produced a panchayat without a majority of the Jat (Farmer) caste in one party. Indar Singh was returned as sarpanch, and one other man of Jat caste was elected as well. There was one man of Harijan caste, a man of Potter caste, and a woman of Goldsmith caste. In 1975, when the village collaborated in International Woman's Year, an additional woman member was elected. The next election was due to take place in August 1978, two days after I left.

In 1978 the party disputes in the village seemed on a personal level to be just as intense as they had been in 1965, but by this time the panchayat had undertaken several important projects that appeared to be detached from party considerations. In these it acted as a single entity in what could properly be called the *public* interest, even though, significantly, its behavior still did not involve use of its legal coercive power.

The clearest symbol of this new activity struck me immediately on the first day of my return trip. On the road to the village there was a freestanding hand pump just where we entered the village lands. It was conspicuous because previously there had been no indication of any kind of the boundary of the village lands, and because in the old sense the pump itself was useless. No particular household benefited by its location. It was too small for agricultural purposes and much too far from the houses for anyone to come to for washing or drinking water. I naturally asked Surindar for an explanation, and he said the panchayat had installed it at the time the road was paved. There was another pump on the other main approach to the village, the road along the canal minor from Uncha Pind. The pumps signify that the people of Shahidpur welcome travelers on the road and offer them this convenience. This of course assumes that the people of the village exist as a collectivity and that the panchayat is its representative, and this is an assumption I didn't see any one act on in public in 1965.

The construction of the village school was similarly an act of the panchayat taken entirely at the initiative of its members acting for the village as a whole and not on the basis of party divisions. The state and national governments are both firmly committed to increasing educational opportunities, but according to the Secretary of the Speaker of the Panjab legislature (who recalled the case), the public demand for schools is so great that the government is forced to choose among requests which in total far exceed available resources.

The main story of the school will be told after more has been said about economics, kinship, religion, and parties. With respect to the division of labor, however, the important point is that in order to obtain the school the panchayat had to present the case to the legislature and to provide the land and building. The government, if satisfied of the bona fides of the presentation and of the need for the school in this area, would then pay the teachers' salaries. The cost of the building was Rs. 45,000. Of this, Rs. 10,000 came from a state legislative fund; about Rs. 20,000 came from large individual gifts, and the rest, about Rs. 15,000, was raised by the panchayat in relatively small amounts from village households. Indar Singh described the collections as "taxes," but they were not the legal hearth tax and the panchayat collected them not by force of law but by persuasion. Payment of the tax, as Indar Singh said, was "voluntary." The land for the school consisted of about 1.2 hectares of the cremation grounds and was given by the panchayat as their legal owner. In 1965 nothing suggested the panchayat taking such an initiative. In effect, what has happened is that the legal panchayat has come to use the old model of a panchayat acting through moral suasion; now, however, this traditional activity has been modified by the new idea that the panchayat represents the village as a political collectivity in a governmental context rather than in the traditional kinship and religious contexts.

Other actions taken by the panchayat reflected its new decision-making autonomy less clearly, but they too have been important. They included raising Rs. 7,000 for paving the road through the village lands (a levy required by the state) in 1969, and several smaller levies to improve village monuments—also "voluntary" payments.

In 1978 the legal powers of the panchayat were still largely held in check by households organized through the idea of parties. But for some purposes, for many villagers, the ideas of a corporate village and an autonomous panchayat acting for it had become important as well.

THE COOPERATIVE CREDIT SOCIETY

Since the early part of this century, village cooperative societies have been advocated by an influential group of high-level administrators and others as the proper source of a large portion of rural credit. They were to be independent village-level institutions controlled by farmers and intended to operate in the farmers' interest, freeing them from the grip of the rural "moneylender." The moneylender in turn was considered something of the villain of the piece, taking advantage of the farmer's gullibility and simplicity, and of natural adversity, to gain control of the resources of village agriculture and weigh it down with a burden of usurious interest. The fact that cooperatives did not work was never blamed on the concept of cooperation as such, but always on the farmer, and in this way their failure was not taken as an indictment of the government that imposed them but as a justification for it.

In 1964 those who warned me of the corruption of panchayats usually also said the same of village cooperatives. It was often said or suggested that the cooperatives were dominated by a few powerful landlords or some other small clique who gave the money only to their friends and relations and did not use it to better the village as a whole.

As with the reports of panchayat corruption, it was inherently unlikely that such a description represented the view of villagers themselves. If cooperatives really did provide an important resource for which there was great need, it seemed highly improbable that *no one* would recognize this fact and defend it within the village. At the same time, as with the panchayat, it seemed equally obvious that the critical assessments could themselves make the cooperatives a sensitive topic within the village. If I inquired too pointedly into the matter I would run the risk of being identified with these viewpoints and with whatever misunderstandings the villagers might see them as demonstrating.

Unlike the panchayat, the village cooperative was never mentioned spontaneously, although the villagers had ample opportunity to do so. Whenever I met new people in the village, American interest rates and credit practices were the most common things I was asked about, after the inevitable inquiries about the assassination of President Kennedy. Yet even when talking directly about agricultural costs and credit, no one brought up cooperatives or cooperative interest rates.

Eventually, when I felt my purpose would not be misunderstood and after I had casually picked up a good sense of the membership of the cooperative governing committee (and at the urging of M. S. Gill), I raised the topic myself when several members of the committee were present.

I described the common accusations. Since there were no large landowners in the village, the idea of their domination was easily dismissed as irrelevant. Some critics saw the village parties as the problem, and I mentioned this. The cooperative committee was clearly divided by party and the dominant party of the panchayat was also dominant here. Indeed some of the members were the same men. It was explained, however, that the law and the regulations of the cooperatives were quite explicit about who would get what loans, and for what. The requirements for record keeping were also quite detailed. I was shown the record. In effect, there was no practical way to give loans only to members of one party or the other. All members were eligible, all members received loans, and people of both parties were members (this I knew independently and directly). The bias came not in giving loans out, I was told, but in collecting the repayments. It was more difficult for people to ask their friends for the return of money and, by definition, people in the same party were friends. Hence while loans were available to all and given to all, the repayments were not demanded so forcefully and consistently from the borrowers in the majority party.

Was this corruption? Yes and no. For some critics, it probably would have been enough to prove that the problem lay with the villagers and not with the system. But from the village point of view, the practices did not amount to the foolish destruction of a basic resource for development, because the cooperatives were not in fact such a resource.

The main sources of credit described in 1965 were, in order of preference, relations, friends, and agricultural jobbers called *arthīs*. Credit from all these sources together, however, was scarce. Commercial banks were oriented exclusively to town businesses.

The *arthīs* had been subject to urban and Western criticism in much the same terms as the cooperatives themselves. They were assumed to charge usurious interest rates and farmers' reliance on them was construed as indicating the same lack of rationality implied in the charge that villagers corrupted the cooperatives (for a definitive and influential expression of this view coupled with an argument to the effect that Indian village farmers needed less total credit than Western farmers, see Darling 1947: especially Chapter X). In fact, however, the *arthīs* earned their pay. They served both as wholesale dealers and general rural bankers. Their offices surrounded market squares where farmers sold their crops. Farmers would borrow money from them against crops and sell their crops through them. If there was cash left after repaying outstanding debts, farmers in turn could lend it to the *arthīs* and would collect substantial interest on it. Since there were many *arthīs*, they competed with each other for the farmers' business. And since there was only one market within a day's travel of Shahidpur, farmers were careful about their credit reputations with the jobbers, so there were checks against abuse on both sides.

Farmers in Shahidpur regularly awaited selling advice from their jobbers. It was in the *arthīs'* interest to give good advice, since when the price was high to the farmer, the *arthīs'* profit per unit was higher as well and there was less risk of having to hold the material for long. It was theoretically possible for an *arthī* to buy at a lower harvesttime price and wait to get the benefit of the later higher price for himself, but it was not practical because of his lack of good storage facilities.

In contrast to that of the *arthīs*, the orientation of the cooperatives was hardly more helpful that that of commercial banks. Since its introduction from Britain in the late nineteenth century, the cooperative movement had been at least as much a program of social reform and education as a way to provide credit for agriculture as such. Cooperatives were intended by their urban proponents to teach thrift; the village bodies were to be "thrift societies" (Hough 1966:55ff.). The members were to learn this thrift by contributing their own savings and then lending them out to whoever among them was in greatest need at any point in time.

Two major problems with the basic conception of this movement stemmed directly from its origins in British socialism. The first was that it was directed at an unreal problem. Teaching Indian peasants thrift is as presumptuous as proposing to teach fish to swim. The second was that the amounts of money involved would have to be small. To use such societies for large-scale infusions of capital to develop agriculture would have defeated their avowed social purpose. The farmers, for their part, had other means to assure security and precisely because of the constant discipline of living in a situation where there was very little that could be diverted from productive use at any point in time, they knew that money was a relatively unfruitful form of savings. Hence they could not and would not put large amounts into the cooperatives. They quite naturally regarded them as a government imposition to be used as efficiently as possible, but not allowed to divert too much effort or too many resources. If money had to be put up by villagers to meet government quotas, it would be—and then it would be lent back out. If the government put in token amounts on its own to stimulate local activity, it too would be lent out. For serious venture capital, for which the villagers felt a great need, they simply had to look elsewhere.

The changes in the activities of the village cooperative between 1965 and 1978 were entirely the result of a radical reversal of official policy implemented through a reorganization and expansion of the national and state cooperative structure. By 1978 about the only link with the past was the basic legal concept of the cooperative itself. A cooperative continued to be a public corporation whose members owned shares in it, and which did business with those same members and shared profits with them. Villagers were shareholders in the village cooperative, on this basis. The village cooperative, in turn, was a shareholder in higher-level units. In fact, it was a shareholder in two such units, and as such the single group of people actually had two distinct legal identities. At the top of the system, in 1978, were six "Apex" bodies. Of these, four were important to villagers. They were: (1) the Apex Cooperative Bank; (2) a federation of cooperatives for marketing and purchasing called Markfed; (3) a federation of dairy-processing and cattle-breeding cooperatives called Milkfed; and (4) the Land Mortgage Bank. The other two organizations, primarily of urban interest, were a federation of labor cooperatives (Laborfed) and a federation of housing cooperatives for underprivileged groups, called Housefed.

Historically, the core of the system was the Apex Cooperative Bank. The Panjab Apex Bank was founded in 1949. Its members were Central Cooperative Banks at the district level, and their "primary units" in turn were the village cooperative banks. In 1978 the other Apex organizations were also members of the Apex Cooperative Bank and thus could obtain credit from it. The state government was also a member and its purchase of shares (not balanced by demands for loans) was a very important source of operating funds for the whole system. The Apex Bank was connected in turn to the Reserve Bank of India and was a major conduit by which Reserve Bank funds were channeled into the state cooperative system. In fact, it was the Reserve Bank that initiated the change in fiscal policy that stimulated the reorganization. Officials of the Reserve Bank had been urging the adoption of a system of crop loan credit (credit secured against future crops) at annual meetings with state officials every spring since 1964, but the state had always rejected it, saying it would not work (farmers would default, etc.). However in 1967, after the state was divided, there was a new Development Commissioner, S. S. Grewal. He immediately prepared to provide a crop loan system out of state resources if necessary. In the spring, when the Reserve Bank renewed its offer, Grewal accepted it. As I understand it, twelve million rupees were initially asked for and granted, and in the first year ten million were actually lent out, mainly in kind in the form of fertilizer. The central government had given the state six months to set up a system. It was done in three, although of course there were continual adjustments thereafter (M. S. Gill, personal communication).

Since 1968 the Central Cooperative Banks, the Land Mortgage Bank, Markfed, and Milkfed, though autonomous financially and administratively, have complemented each other and form a single system when seen from the village perspective. The crop loans, which are short-term loans in the nature of the case, are given by the Central Cooperative Bank through the village cooperative as its Primary Bank. Crop loans in 1978 were limited in amount to Rs. 9,000 per applicant, and were divided according to intended use. One component (Rs. 1,500) was for pre-sowing operations, a second (Rs. 6,000) for fertilizer and other inputs to the growing crop, and a third (Rs. 1,500) for harvesting. The policy was that the second component should always be supplied in kind. Applications for such loans were approved by the local committees, and action was prompt. If past loans were not repaid to the regional bank, the local society was not given new material.

The Land Mortgage Bank has the responsibility of providing credit for major capital purchases or equipment. Loans are secured against land. Again, the village society is the "Primary Bank" and applications are initially submitted to it, but from this point the procedure becomes much more complex. Repayment is made by the borrower to a designated regional banking office. In 1978 this was in Morinda. At that time, loans for tractors could be for amounts up to Rs. 60,000, exclusive of accessories. Borrowers were required to own a minimum of ten acres (4.04 hectares) of land. Repayment could take up to twenty years.

Milkfed supports dairying and loans are repaid out of proceeds, as with crop loans, but the terms are naturally longer. Application and repayment do not involve the village cooperative, so they need not be described here.

Markfed is not a credit society and the village cooperative is not a member of it, but at the state level it has the responsibility of purchasing and distributing the fertilizer and other materials that the local body receives from the Central Cooperative Bank. Markfed makes its purchases on the open market and stockpiles it, passing to farmers the advantage of mass purchasing. On the other side of the balance, Markfed buys major foodgrains for central government stockpiles from individual farmers in the market at publicly announced floor prices (as do two other purchasing cooperatives, on alternate days). Farmers remain free to sell for higher prices if they can get them. The floor prices are calculated to provide the farmers a reasonable return over their investments. Farmers generally agree that they do, and higher prices are usually not available. Markfed thus provides essential support that keeps the new cooperative credit system from declining to the status of the old.

The increase in new fertilizer loans from Rs. 14,000 in 1969 to Rs. 70,505 in 1978 gives the best measure of the increased importance of the village cooperative as the Primary Bank in the crop loan system. The increase in credit from the Land Mortgage Bank similarly indicates the increasing importance of the village body as *its* primary unit. In 1965 there were no such loans. In fact it appears that there was only one Land Mortgage Bank in the entire state (cf. Statistical Abstract of Panjab, 1960:261). In 1974 (the latest year represented in the land record which lists liens against fields) loans outstanding totaled Rs. 56,000.

In short, by 1978 the cooperatives had become an established part of the framework of farm management, and the local cooperative was respected as part of that framework.

HOUSEHOLDS: THE MAIN CONCEPTUAL CATEGORIES

Important as the panchayat and the cooperatives have become, households remain the fundamental managerial units of the village.

The normal village term for a household in the division of labor is *parivār*. The head of the household is the *mālik* ("master"). Adult males and others with recognized rights over its property are *hakdārs* ("rights-holders"). It is understood, however, that all in a household are involved in its decisions in some way. Further, it is also understood that the resources that must be managed include not only material objects but certain sanctioned concepts. These specifically include certain *legal* concepts of the family and certain kinship concepts that under some circumstances can be enforced as law. Strictly speaking, these definitions should be given in the section on basic kinship concepts, but it will be clearer to anticipate that discussion somewhat here.

The main legal concept that a court could apply to most village households is that of the Hindu joint family, which is defined in several laws (not always mutually consistent) pertaining to marriage, succession, rights of children, rights of women, and landholding. The only village families that would not be construed as Hindu joint families in a court of law would be three Muslim households, all Cotton-ginner by *jātī*.

The Hindu joint family is constituted by a distinctive rule of succession. There are two such rules, both recognized as equally "Hindu." In Shahidpur, as in almost all of India, the Mitakshara rule applies. This pertains both to the division of property and to its reversion. The rule has four major elements. The first is that only men have hereditary rights of ownership. The second is that they obtain these rights from birth. The third is that they hold these rights equally with all brothers and with their father. The fourth is that rights of sons preempt (take precedence over) rights of brothers. This applies to all "ancestral property," but not to property that is "self-acquired." Any property obtained by inheritance is ancestral, as is any property obtained with inherited resources.

The rule means, for example, that if there are three sons of a father named

A, each of the sons will own one-fourth of the father's original estate, from birth. The father will own the remaining fourth. If one son, *B*, should die, his share would revert to the father and the remaining sons (his brothers), with the result that his brothers and father would each own a third. If one of the remaining sons, say *C*, had a son of his own, *D*, that son would become half-owner of *his* father's (*C*'s) estate, but would have no direct claim on the shares of *C*'s brothers or father. Then if *C* himself died, his remaining half-share would go to his son only, making *D* a third-owner of the original estate. If the original father or the other brother died, *D* would also have a claim on the share that would have fallen to his father (*C*).

The rights include both rights of ownership and rights of preemption. Preemption means that if the original father (*A*), for example, had sold the ancestral property to anyone but his sons without their permission, they could reclaim it at any time by paying the buyer the original sale price. The sons would have the same right even if the father sold only his "own" portion. If the sons did not exercise this right, the father's brothers could. If the brothers did, the sons could later claim the property from them in turn on the same basis.

Sons, in effect, lock property into one's own descent line. They make it difficult for one to sell one's own ancestral property, but by the same token they make it extremely difficult for anyone else to attempt to take it away.

There are many variations in the interpretation of this basic rule—such as, for example, those concerning the force of the son's son's (*D*'s) claim against a widow of one of his father's other bothers, or as against that of the original father's brother in the event of the death of the father (*A*) and two remaining sons without issue. These vary by region and, often, ethnic group and *jātī*, and such variations make up much of the substance of Indian "customary law" insofar as they are recognized and enforced by courts.

In Shahidpur, traditional rights of preemption are said to have extended much further than modern law recognizes, and this extension still has a certain moral force. More important, the local operation of the Mitakshara rule is influenced by local translations of the key technical ideas, and among most landowners it is further interpreted by Jat customary law. Jat customary law, also enforced by the courts, does not allow the sons the right to demand a division of the ancestral property against the father's wishes. If the father opposes division, the property can only be divided after his death. The legal birthright still exists, however, so the father cannot by any means treat the property as if it were his own in a Western sense.

The legal term "ancestral property" corresponds, in village discussions, to the Panjabi phrase *jadī jaidād*, which means simply "old property." Villagers who explained this phrase to me took it rather literally and did not bother with the legal test that such property must have been inherited itself or acquired out of resources that were. By the same token, the legal term "self-acquired" was rendered in the village simply as "new" property. In practice villagers are more inclined than the law to make the presumption that property is ancestral.

The English phrase "Hindu joint family" is not usually used in the village. The term *parivār* is usually employed in its place. When I first began systematic elicitation of kinship concepts, this was defined for me as "a group of relatives who cooperate in some property." It was distinguished from *ristādarī*, the group of all of one's relatives. Very distant relatives would be in one's *parivār* if they cooperated in property; very close relatives would not be in it if they did not cooperate.

A legal joint family of record might in fact be divided into several distinct village *parivārs* living in separate houses, although villagers would recognize this larger group and have no trouble speaking of it as a *parivār* as well.

A functioning household (*parivār*) is expected to consist of one male line and associated dependents. The male line would be the master (*mālik*), his wife and children, surviving parents, and perhaps an unmarried brother or two. If children include two or more sons, the family will stay together until both of the sons have sons of their own or until there is a major dispute. Adult sons would generally be described as *hakdārs*, rightsholders, although women with clear rights (for example, widows holding property in trust) could be too. While it is understood that the *mālik* predominates, it is also understood that the rightsholders all exercise an element of independence in judgment and that in the case of major disputes, they represent the potential lines of cleavage along which new *parivārs* might be formed. In any case, if two lines of sons descending from sons seem well established, the expectation is that the household will divide. The stereotyped occasion of the division is that the two mothers will quarrel because each will seek advantages for her own son.

In 1965 only one family (with a resident daughter and her husband, but no natural son) had members who would not be expected under the stereotyped description. In 1978 the pattern still held. There had been no change in the basic concepts or the basic expected organization itself, although demographic changes had somewhat changed its significance.

HOUSEHOLDS: OCCUPATIONAL PATTERNS

There were 118 *parivārs* in the village in 1965 and 153 in 1978. Each villager knew most of the members of most of the households personally, so that in mature considerations it was possible to take into account the character of the members and their resources directly and precisely. This much information would not be useful here, however, so as a second-best system I will follow the villagers and construct descriptions by grouping the households under three major headings: *jātī* ("caste": hereditary occupational group), *got* ("clan"), and parties. Because of certain very prevalent misunderstandings, I should explicitly say that this is in no way to be construed as a "Jajmani system" as described in the literature summarized not too long ago by Beidelman (1959), or even in the more-attenuated form recently argued for by Breman (1974). Such accounts, which suggest that the system of occupations is based on—caused by—caste and enforced as a system of caste prerogatives, are uniformly based on incomplete and inappropriate data. More specifically, they involve taking the way of talking about occupations by using caste labels that I am going to replicate here as evidence that caste is the basis of occupations. Instead, occupations are obtained on the basis of whatever resources people have—education, capital, and skills. Caste ideas, which are appropriate for reasons that will be given in Chapter 6, are applied only roughly and after the fact.

There are eleven recognized *jātī* groups, divided into *gots* and crosscut by parties. The *jātīs* are arranged in a rough socioeconomic ranking. This ranking, along with some of the most general features of household size and structure, are given in Tables 4.1 and 4.2.

Table 4.1 shows, among other things, that some important differences in household size and composition between the different *jātīs* in 1965 had been sharply reduced by 1978. The summary row gives the main point: while the population of the village has increased 21 percent, the number of distinct families has increased 29.7 percent. Families have thus become smaller and more numerous, but the trend in this direction has not been the same in all *jātīs*. By far the largest groups in both periods were Jat and Harijan, together making up about 70 percent of the village population. In 1965 the average Jat family was 26 percent larger than the average Harijan family—a sharp difference. The greater size was associated with greater generational depth, so that it was quite accurate to say that the Jats as a group were more highly organized than the Harijans. In 1978 the average Jat family was just 9 percent larger. While Jat families had become smaller, Harijan families had remained virtually the same size. As the table shows, there was a general leveling out of family sizes across all the larger groups.

Table 4.2 shows the changes in house size as indicated by the number of housing units per house. A "unit" means a discrete area, which may be an open work area or a building with from one to eight or nine rooms. Houses with more units generally have more and larger rooms and more work space.

As the table indicates, Jats continued to have the largest houses of any major group in the village, with an average of 1.9 units each. Harijan house

Table 4.1

Household Population by* Jātī, *1965 and 1978 (Residents Only)

Jātī	*Population of* jātī		*Number of families*		*Average family size*	
	1965	*1978*	*1965*	*1978*	*1965*	*1978*
Brahmin	24	19	3	4	8.00	4.75
Jat	332	373	45	57	7.38	6.54
Mason	42	60	7	9	6.00	6.67
Goldsmith	16	13	4	5	4.00	2.60
Barber	2	8	1	1	2.00	8.00
Water-carrier	2	1	1	1	2.00	1.00
Potter	78	98	5	11	15.60	8.91
Cotton-ginner	10	12	1	3	10.00	4.00
Sadhu	10	16	1	1	10.00	16.00
Harijan	242	310	41	52	5.90	5.96
Sweeper	48	67	9	9	5.33	7.44
Total	806	977(+21.0%)	118	153(+29.7%)	6.83	6.39(−6.5%)

Table 4.2

Units of Housing per Household (by Jātī)

Jātī	*Single*	*Double*	*Triple or more*	*Unit houses*	*Total houses*	*Total units*	*Total new or remodeled units since 1965*	*Average units per house*	
								1965	*1978*
Brahmin	0	3	1	0	4	9	4	1.7	2.3
Jat	24	25	12[a]		61	116	35	1.9	1.9
Mason	7	1	1	0	9	12	2	1.3	1.3
Goldsmith	5	0	0	0	5	5	2	1.0	1.0
Barber	0	1	0	0	1	2	0	2.0	2.0
Water-carrier	1	0	0	0	1	1	0	1.0	1.0
Potter	5	4	2	0	11	19	9	1.4	1.7
Cotton-ginner	3	0	0	0	3	3	2	1.0	1.0
Sadhu	0	1	0	0	1	2	1	1.0	2.0
Harijan	38	13	3		54	73	20	1.2	1.4
Sweeper	6	4	0	0	10	14	6	1.1	1.6
Total	89	52	19	0	160	256	81	1.4	1.6

[a] Three of four units; one of six units.

sizes had increased, from 1.2 units per house in 1965 to 1.4 units in 1978, and the gap in size between the houses in the two groups had narrowed, but Harijan houses were still considerably smaller (about 36 percent fewer units, but actually less than half the size under roofs on the average). Improvements were greatest among the Jats, but not by a great deal: 30 percent of the Jat houses had new or remodeled units between 1965 and 1978, while for Harijans the proportion was 27 percent.

Of course size is not the only aspect of housing that is important; there have also been important improvements in quality. These were more conspicuous among Harijans than among Jats, with full concrete floors, screens on windows, and even glass jalousies. Among the Jats, changes were usually more strictly functional, like floors on areas for buffaloes, elimination of barns for families whose occupations had changed, and needed repairs.

Other changes were related to occupational shifts. Obviously, the trends are not simple, and the closer one looks at them the more one moves from the seemingly detached generalities to the interplay of the specific individual strategies that underlie them and compose them. The best way to see this is to turn to the review *jātī* by *jātī*.

Brahmin

The Brahmin's stereotypic occupation is that of hereditary priest, serving especially as family religious teacher and priest to Hindu kings. There are no kings in Shahidpur and none of the Brahmins earn money from their traditional calling in any sense whatever. The three families of Brahmin *jātī* in Shahidpur in 1965 were all of separate *gots*. At the time, the family that could be most readily identified with *any* single occupation was Babu Ram's. He had established his small two-room shop when he came to the village about ten years before. Babu Ram and his family lived partly in the shop and partly in an old house nearby that belonged to Indar Singh. In the winter of 1965/66 he began to build a house of his own, just inside and to the east of the village gate.

The second family was headed by Banarsi Das, who served as village doctor. He had been in the village only a few years, but had built a new house next to the village gate, facing outward. He used the bare and rundown office of the village panchayat as a clinic and his pharmacopia consisted mainly of a few old-fashioned antiseptics. He did not have formal medical training and his practice could be best described as casual. Some villagers went to him for minor problems; many lacked confidence in him.

The third family was that of Alekh Ram. He was the hereditary owner of the centuries-old Hindu temple near the southern entry to the village *abādī* (near his modest-sized, but quite comfortable, house), in which he said he prayed every morning "for the welfare of the village." He was a retired schoolteacher as well as village postmaster. He also did some farming, with a small amount of land and a share in a well, and he was secretary of the village cooperative society.

In 1978 Babu Ram had been "retired" for about two years, forced to give up the shop by poor health. He was living in the house he had completed, two rooms and a courtyard.

The Post Office was also closed. About halfway through my visit, Alekh Ram returned from having serious abdominal surgery. His son had been a teacher living and working outside the village in 1965 and was still doing so in 1978. They still held some land, but during my brief visit to his house the day Alekh Ram returned from the hospital, I thought it would seem disrespectful or unfeeling to question him on the details of the arrangement on which it was being farmed. Except for a large electric fan (and electrification) and the fact of being able to afford surgery, Alekh Ram seemed no better off than he had been in 1965.

Banarsi Das, by contrast, appeared to be prospering. He had constructed a special building to serve as a clinic with a large waiting room in front and an examining room in the rear. The waiting room was equipped with an overhead fan, a desk, a bench, and a glass-fronted cabinet with a substantial stock of medicines, not only antiseptics and the like but antibiotics and other sophisticated and specialized drugs to be administered both orally and by injection. He had clearly improved his range of knowledge and skill, now kept regular office hours, and commanded a much wider following than before. I was in his waiting room twice that summer. It was busy both times, and some of the complaints were moderately serious. It was obvious that those waiting respected his skill, and equally obvious that he was attentive and serious in his concern for them. He had also invested in land, mainly in mortgages, and had by then an income equivalent to that from about three hectares. His son owned one of the village motorcycles.

The fourth Brahmin family reported in 1978 was entirely new to the village. Its head was another doctor, who reportedly had some surgical training. He had come in 1968 and had built a new house on the north side of the *phirmī*, across from the village mill. He traveled daily to an office in a nearby village. His *got* was different from that of any of the other Brahmin families.

Jat

Strictly speaking, "Jat" does not name an occupation. The term corresponds to Western ethnic group names like "Scythian." (It is said by some that Jats are descendants of the Scythians.) In the few discussions in Hindu literature that specifically mention Jats, there is a wide range of opinion on where they might fit into the "caste system," but the predominant view seems to be that they are not in it at all. Other discussions place them at the second level, with Rajputs. Either way it is of little concern to the Jats, although when I asked most said they were "like Rajputs." It is generally assumed in Panjab folklore and popular discussion that Jats own land and farm it.

In 1965, sixty-seven Jat families out of seventy-three (including nonresidents) obtained income from farming. Sixty-four actually owned land. The average size of the farmed holdings was 1.97 hectares and the range extended from a few square meters to 9.55 hectares. In 1978 the number of resident families obtaining income from agricultural land had declined to sixty-three out of seventy-six. Fifty-seven Jat families were owners. The average size of a holding had increased to 2.31 hectares and the range extended from the same minimum to 9.70 hectares. It is harder to obtain figures for the commercial farms proper in both periods, since there is no single criterion by which they can be identified. The general estimate in 1965 was that about fifty families were involved in farming as a major occupation: two Harijan families, one Sadhu family, and the rest Jat. In 1978 the estimate was about thirty families: one Sadhu and the rest Jat.

In addition to land, in 1978 thirty-two resident Jat families owned draft animals (one or two "plows," not more), fifty-six owned buffaloes, thirty-seven owned diesel or electric motors (usually one; five families had two), eight owned oxcarts of the old type and twenty-four owned oxcarts of the new type. The second motorcycle was owned by the same Jat family that owned the ten-ton truck. This family had no land and farmed a small amount, less than a third of a hectare. One son was away in the Army. Another family, whose head was an officer in the Indian Army and thus lived away from his family (his brother managed the household), owned the one automobile in the village, which was kept in a garage built for it in a yard outside the *phirmī* that was shared with a small dairy. This family owned a substantial amount of land, which was worked on a rental arrangement by Indar Singh, who was a close relative (son of brother of father) and good friend. Three Jat families owned tractors.

Most Jats who have left farming have done so individually rather than as entire families all at once. They have usually mortgaged or sold their land to those who have stayed, usually close kin. In a few cases, they have simply left their land "in the care of" (a legal category, recorded in the land records) another. Mortgaged land and land in the care of another can be reclaimed. Sold land cannot. Most who have left have taken salaried work. Even among those who have stayed in farming in a broad sense, there has been some recognized diversification: mainly between those who have concentrated on field crops and those who have concentrated on dairying. The result, all in all, is considerably increased occupational differentiation. At least three physical houses of Jat families have, in consequence, been modified to exclude barns and storage facilities, thereby becoming more specialized as residences. Jats who have new occupations, in addition to those already mentioned (billing clerk and cooperative quality control officer), include a public health worker living in Chandigarh (the daughter of Mohan Singh), a member of the staff of an agricultural experimental farm near New Delhi, an employee of a government hydroelectric project, the owner-manager of residential properties in Chandigarh (a family that has moved out entirely), and a sailor. There are also several teachers (at least four) and men pursuing military careers at many rank levels, although these are not new careers since 1964. A few families have moved to other villages to set up new farms there. All of these still have close relatives in the village and were reported as resident or nonresident family members by them.

In 1965 all of the Jat families but one were of a single *got* and considered themselves to be descended from the founder of the village, although no one could trace all the links. The one family of a different *got* was "of" another village nearby and was said not to own land in Shahidpur (on the record he did own a little). The head worked mainly as a partner and tenant of the farmer who later obtained the residential properties in Chandigarh, but who was then one of the largest village landholders. In 1978 there were three new families, all involved in farming and all of different *gots*, which brought the total number of *gots* to five.

Jat households are recognized as the core of the party alliances in the village and were as sharply divided in 1978 as in 1965. I could not find as many households involved in the later period, but feel certain this was only because I wasn't there long enough to attend the full range of events where the divisions would have shown themselves, as will be explained in Chapter 8.

Mason

The families of Mason *jātī* are all of one *got* and in fact of one *parivār* in a broad sense. Although they lived in nine separate houses in 1965 and twelve in 1978, they all were connected by known genealogical links to a common ancestor. In 1965 the men of Mason *jātī* actually answered to the names of three traditional *jātīs*: Mason, Carpenter, and Blacksmith. One family had the tools for smithing and at least two had the tools for masonry and carpentry. Whenever anyone needed help in any of the three trades, he could call on at least one other Mason for it.

The Carpenter's traditional occupation includes making furniture and making and repairing tools (houses are not constructed of wood). The Mason's traditional occupation is brick and stone construction of all kinds and, by extension, construction in cement and iron girders. Blacksmith's work was metalwork, especially the making and repairing of iron tools. The village Masons did all of these types of work in fact (but not on a "traditional" payment basis) and in addition were village millers and operators of several other types of powered machinery, as noted. Two families also obtained some income from land, just under 0.11 hectares in all. Several men worked more or less regularly in nearby villages. Two, one with his family, were doing construction work in New Delhi.

In 1978 the situation was basically the same, except that everyone was busier and the work was on a larger scale. One of the Masons who had been in New Delhi had returned and built a new house on the urban pattern, between two brothers on the north side of the *phirmī*. He joined one of the brothers to open a "wedding furniture shop and workshop" in Uncha Pind, to which they commuted. The brother owned another of the motorcycles. An additional diesel had been obtained and an electric motor as well. One diesel was in the shop in Uncha Pind. A family head who had been in the village doing masonry and carpentry in 1965 had gone to work in Germany, although his family stayed behind. No one still held land.

Masons avoided party disputes in both periods. They arranged their occupations among themselves so that each family, or at most a pair of families, had a virtual monopoly on some particular set of activities in the village. Under these circumstances the only effect of party alliance would have been to alienate about half of the families that would otherwise have been customers.

Goldsmith

A Goldsmith traditionally works gold and makes jewelry. In some of the scholarly literature he is represented as having a derivative function as a moneylender, but this was never suggested in the village. In 1965 no one had goldsmithing tools. When I asked why, I was told with a laugh that there wasn't any gold in the village to smith. One household head was the village "master tailor" (and did very good work on both men's and women's clothes of the traditional patterns). Another did some tailoring part time. A third owned a metal box for transporting eggs fastened to the back of a bicycle which he used in a small brokerage and delivery business. The rest of their income, which was not great, came from agricultural labor. None were reported in parties.

In 1978 the main change was that the master tailor had been able to build himself a new house of three rooms, including a new tailor shop. The woman of Goldsmith *jātī* on the panchayat was the wife of the family head who had done occasional tailoring but mainly agricultural labor. Her position and his occupation together suggest that he had moved from being neutral in party matters to being a supporter of the dominant group. The owner of the egg crate still had it, but had gone somewhat mad from the strain of many personal difficulties. He was quite poor. Goldsmiths had no land at either time.

Barber

In 1965 the one Barber family was not wealthy but comfortable and well regarded. The traditional job of a Barber is shaving people and cutting hair; his stereotyped "special equipment" is a razor. Sikhs in the village, including the Barber, generally made a point of following the injunction of tenth Sikh Guru, Guru Gobind Singh, which was to never let a razor touch their hair or beard. Obviously the Barber did not follow his caste calling. Instead, as a general master of ceremonies in the village, he was the formal go-between in weddings for families on the Jat side and supervised the ritual performances as well. His wife had a formal role in weddings as his counterpart in a series of exclusively women's rituals and was also the village midwife. Their income came from these activities, from a son who was a teacher living out of the village, and from some agricultural field work. They had no income from land of their own and were neutral in party matters. Or rather, since the Barber was privy to almost all gossip and was a genial and cheerful gentleman on good terms with virtually everyone, he was described as being in "both" groups. The Barber died a few years after my first study and his son moved back into the village. There has been no notable change in their fortunes; the house in 1978 was also unchanged.

Water-carrier

The Water-carrier *jātī* was represented in 1965 by a widow and her son (her only child). In 1965 the son was just graduating from the nearby higher secondary school and was awaiting word of his acceptance into the Air Force. Their life in the village had been hard. Traditionally a Water-carrier would carry water from the main village well to the various houses for cooking, washing, and the like. However almost all the houses had hand pumps and those who could not afford one also could not afford a Water-carrier. Hence the family's main income came from agricultural labor, supplemented by considerations for a small traditional role the mother was sometimes asked to take in weddings. The agricultural work (such as cutting and trimming sugar cane) was difficult for a rather petite woman. The son was husky, hard working, and very serious. He could assume more difficult (and hence more remunerative) tasks and work longer hours, but his capacities were severely strained by trying to balance these tasks against the pressing need to do well in school. Military service required high school graduation.

He was accepted in the Air Force a few months before I left, and I met him once after that when he was home on leave after basic training. He was still serving in 1978, and by then had a family of his own. His mother was still in the village. He was doing well and apparently making the Air Force his career.

Potter

The Potters are also all of one *got*, although their mutual relations cannot be traced. The stereotypic work of Potters is making clay cooking and storage vessels of all kinds for village families. As a derivative of this, Potters are also expected to operate brick kilns and to haul dense materials on mules and donkeys, which are part of their "special equipment." In 1978 as in 1965, two of the Potter families actually made and distributed pottery. For this they owned two kilns, two hand-powered potting wheels, and the mules and donkeys noted in the previous chapter. In 1978 they had an oxcart of the new type; I am not sure if they had a cart in 1965. The pottery was provided to other village households on an annual contract basis, with the charges reflecting the size of the families. However, the range of types of pottery needed has narrowed consistently as brassware and other types of materials have become more common for cooking, as electricity has eliminated the need for clay oil lamps, and as the hand pumps have reduced the need for water storage jars. Accordingly, the working potters have become more specialized in the few things that are still in demand and have begun to sell what they make more and more widely. In 1978 they were making only a few kinds of pieces almost as forms of folk art. They were considered well known for them and sold them over a wide area.

The Potter families that do not deal in pottery draw their income from two main sources: agricultural work and military service. In 1965 two families also obtained some income from land, a total of 0.26 hectares. By 1978, none had rights in land of their own, but several worked for farmers on full-time bases. One young man, who worked for Indar Singh, took part in most of our conversations. He showed himself to be extremely knowledgeable about all aspects of crop management and was obviously well regarded by Indar Singh's family. A good part of the increase in the population of this group of households is due to the return of three families who had been reported as out of the village in 1965 on military duty. In 1978 they were back and drawing pensions. About five men were away.

Potters who are involved in agriculture also tend to be involved in the party alliances. The Potter who is on the panchayat works for farmers and supports the dominant party.

Although the tool inventories of the Potters have not changed and remain rather simple, their houses show substantial improvement, apparently reflecting both their military incomes and increased agricultural income. There has been substantial new construction and much trading of units back and forth. In 1978 only four families were living in the same houses they had lived in in 1965; seven families had either added units, built wholly new units, or shifted into units that were improvements over what they had before.

Cotton-ginners

The Cotton-ginners are all Muslim by religion and are of one *got*. They are long-term residents of the village, remaining undisturbed during partition. In 1965 there was a father and three adult sons, each with a family of his own. They owned simple, hand-powered ginning devices and together kept a circuit of villages they visited in their work, which was in fact that of their *jātī* name. It included ginning the cotton that farmers used for household fiber and seed. Payment was by weight in grain corresponding to the weight of cotton ginned. Since cotton for market was not ginned the income from this source could not have been great. The second major part of the Ginner's stereotyped work was (and is) that of restoring the loft to the cotton batting which fills the comforters and mattresses that are used in the winter. Ideally, each family will have one of each per person and perhaps a spare or two and these will be reconditioned every other year.

The father died in 1972 and the three houses of 1978 were those of the three brothers. A diesel-powered gin had been added to the tool inventory. According to one of the brothers, they used this for larger-scale, wholesale work. (The brother who operated it had it out of the village when I stopped by, and I could not tell from what was said exactly what the engine did.) In addition, one household head also worked as partner to a farmer. This means he helped with all agricultural work and in exchange took a share of the crop—in this case a sixth. The Cotton-ginners live in substantially the same poor and crowded quarters as previously, having only traded some rooms among themselves.

Sadhu

A Sadhu, stereotypically, is a wandering mendicant who does not have an occupation in the usual sense. He is rather a person who has given up his occupation for a life of contemplation. Moreover, since Sadhus are generally thought of as celibate, or at least as without families, they cannot pass this particular activity on through inheritance. Obviously, this is not what the term means in the village system of management.

In the present context, the name refers to two different but related things. First, it refers to the fact that a major block of land in the village is actually owned by the Udasi order of Sadhus. Udasis are a celibate order with a long historical association with the Sikh religious community. The village dharmsala is maintained by income from this land for the members of the order when and if they visit. There is one resident member of the order, the Mahant ("priest"). The land is cared for by this Mahant as trustee, but he himself does not do the physical work. That is done by his brother, who is not himself an Udasi, and this brother's family.

The second and most important thing the term "Sadhu" refers to in the village is this whole group. No one has ever assigned them any other *jātī* label but Sadhu, although in 1965 it was generally said that they were Hindu, not Sikh, by religion. The one resident son styled himself as a Sikh, however, and in 1978 seemed to be more thoroughly regarded as such. The Sadhus' holdings amount to just over 10 hectares, which keeps them quite fully employed. In addition to the land itself, they had a substantial set of tools: an oxcart, an electric motor, a diesel engine, a sugar defecator, and a tractor with utility wagon. They were quite well off and members of the family have consistently taken part in party alliances.

Harijan

Harijans, along with Sweepers, occupy houses in a distinct section of the village with a separate entrance and a separate (ancient) well. The term "Harijan" is not a traditional *jātī* name and has no occupational sense. It literally means "born of God" and was coined by Mahatma Gandhi and intended as a euphemism for the pejorative term "untouchable" (the English term, which also designates no occupation). In 1965, in Shahidpur, the term was itself considered a bit insulting. It sounded more like a polite way of saying "bastard" than a polite way of saying "untouchable." Either of two traditional *jātī* names was preferred for direct address: Weaver or Leatherworker. The traditional work of Weavers is weaving cloth of all kinds. The traditional work of Leatherworkers is collecting and curing hides and making and repairing shoes.

As in the case of the three *jātī* names for the Masons, it was unanimously recognized that Weaver and Leatherworker were in principle two distinct *jātīs* and that if a person was in one, he could not logically be in another. Yet there was no question that Weavers and Leatherworkers were the same group, whatever one might consider that group to be. At one point, in order to test my assumption that these stereotypes were just that, general matters of belief and not actual descriptions of occupations, I pressed the matter when I found two brothers before me, one of whom said he was Weaver, the other Leatherworker. I went through all the definitions, and they granted them. Then I pointed out their own identifications (each had been asked separately, just previously). The one who had said he was Weaver thought a moment and then simply pointed at the other and said, "Oh, he is Weaver." Obviously the concepts are firm, but applications are maleable. The final explanation was that the label chosen "depended on work." If one wanted weaving done, one found the person one wanted to do the work and called him, naturally, "Weaver." If one wanted shoes repaired, one found the village cobbler and called him, naturally, "Leatherworker." For agricultural labor, one called upon (or answered as) a Leatherworker.

By 1978 ideas had changed to the extent that the term "Harijan" was considered appropriate to use within the hearing of those described and perhaps was even slightly more polite than the occupational terms. It was still not used in address, however. The change apparently reflects greater acceptance of the political assumptions behind the original usage as well as the greater economic leverage (due to the greater scarcity) of agricultural laborers.

In 1965 the Harijan *jātī* contained families whose men were of four *gots*, though two of these were much larger than the other two. In 1978 one additional family of a new *got* was added. The village sarpanch in 1965 was of the second-largest *got*.

Among Harijans, one aspect of the increasing professionalization of village work has been a sharpened contrast between those households whose income actually comes from weaving and those who rely on full-time agricultural work. In 1965 these groups overlapped greatly and all people tended to take whatever work they could get. There were then nineteen looms on the Harijan side. Three of these were *khādīs*, large, framed looms capable of making continuous bolts of cloth about one and a half meters wide. Users of these looms bought the raw cotton locally and sold the cloth, usually outside the village. They were true professional weavers and generally did not do agricultural labor. The other looms were *kumbals*, much less elaborate and less expensive affairs lacking a fixed frame and capable of making at one time only a single piece of cloth of a fixed size, just under a meter wide and a little over two meters long, itself called a *kumbal*. Two *kumbals* stitched together lengthwise formed a standard type of cloth used widely as a wrap, bedspread, blanket, or just about anything else that one might think of. Most of the owners of *kumbals* used them only on a part-time basis, to supplement income obtained on other bases (usually agricultural labor). Farmers would bring them cotton, and they would weave this cotton into *kumbals* for those farmers for a set fee. Those without looms generally did agricultural labor, although this was highly seasonal. When demand was low, they sought work as construction laborers for Masons or on such government projects as roads and canals. I estimated that most families were employed only between one-third and one-half of the time; farmers estimated that the net worth of Harijan houses on the average was about one-tenth that of Jat houses.

In 1978, the number of *khādīs* had increased to six and the number of *kumbals* had gone down to one. All the owners, including the owner of the *kumbal*, were full-time professional weavers. The owner of the *kumbal* was in fact a craftsman of considerable repute who did fine work using only natural dyes and working his fiber in the most careful way. Like the owners of *khādīs*, he generally bought his cotton locally and sold the finished products retail. For the most part, the families that have abandoned weaving have gone into full-time field labor.

As part-time weaving has apparently been abandoned, so has part-time farming. In 1965 five Harijan families obtained income by farming land rather than by working as daily or annually paid laborers. One of these families held its land (0.52 hectares) by outright ownership. The others held theirs on mortgage, rental, and partnership. The total income was the equivalent of the production of 3.68 hectares. By 1978, the number of owners had increased, but not the amounts owned—six owners held a total of 0.72 hectares. Of this, one family owned 0.71 hectares, the only amount even approaching a holding that could provide enough to market. They farmed the entire amount. The other families held plots as small as 0.01 *bīghās* (0.0008 hectares), mainly for vegetables and green fodder for buffaloes.

In 1978 Harijans reportedly had a 25 percent reservation of jobs in government service, which provided a substantial incentive to obtain an education and move out of the village employment system. Among salaried workers reported as residents in 1978 there were six men in the Army, one man who was working in a fruit orchard in Pinjore (a well-known Moghul garden in the Himalayas above Chandigarh, run as both a public park and a demonstration fruit farm), and one who was a police constable in Chandigarh. Another household head, Master Kushhal Singh, lived in Shahidpur but commuted to a nearby village where he taught in a primary school. From clothing styles and conversations, I had the impression that there were at least several other men similarly living in the village and working outside in the area, but the census could not identify and describe them because it asked only about residence and not about employment as such (this would not have been a clear question unless it was broken down into a very long set of specific issues). Kushhal Singh had built one of the new houses on the lane just north of the traditional Jat side, and it was he who first chided me for not asking more about education on the census form, stressing its importance to the progress in the village since my first study. By then, however, all but about eight forms had been completed and even though I and a substantial group around us at the *chauk* by the mill all agreed with him, there was not much that could be done. And there was no way to go back to collect the necessary baseline information for 1965. In 1965 the census had recorded just three Harijan men out of the village in salaried work, all in the Army.

Major tools owned by Harijan families in 1978 included two older-type oxcarts, one electric motor, and one diesel motor. No one owned leather-working equipment. In 1965 just one man of Harijan *jātī* owned such equipment, the only set in the village. Upon entry into the Army, he had been trained as a cobbler (evidently the British took his *jātī* title more descriptively than villagers do) and worked at it during his whole career. On retirement he had been allowed to keep the tools of his trade. He died in 1977 and since that time no one in the village has taken up the work. In 1978 shoes generally came from retail merchants in nearby towns. They were in the form of the older leather slippers, but made of molded plastic.

The Harijans' withdrawal from farming and their abandonment of smaller looms in order to take up farm wage labor might be construed as evidence that their economic position actually declined between 1965 and 1978, especially if one thinks of farm laborers on the model of the migrant agricultural laborers in North America. But these villagers are not migrants. They live in substantial houses, they work in the same place they live, and because of this they have the same access to education and other regional resources as other villagers. They also can obtain much of their food and goods at village costs and/or at the costs of production, which are substantially below the retail market prices that city workers pay. The withdrawal from farming, given all this, is due to the fact that it actually makes sense to concentrate on a combination of wage work and education in order to obtain still-better wage work. The most dramatic evidence for this will be given in the discussion of wages in the next chapter, but here it is appropriate to describe a little more about the improvement in housing stock suggested by Table 4.2.

Although a smaller percentage of Harijan houses than Jat houses were remodeled, the new Harijan construction still raised the number of housing units per family by 15 percent, from 1.2 to 1.4, and produced a substantial improvement in housing quality (many new concrete floors, for example). In addition, the streets on the "Harijan side" of the *abādī* have gone from being about 80 percent paved to 100 percent (even though several entire new streets have been added) and, as noted, a moderate-sized rest house is under construction just across the canal from the Harijan side entrance to accommodate wedding parties and other guests of Harijan families.

Sweeper

The last *jātī* group is Sweeper. All the Sweeper men are of one *got*. In 1965 they were the only group about whose position in the *jātī* ranking there was no disagreement: they were on the bottom (it should be noted, however, that this was not a matter of great seriousness). The reason lies in the name itself. The traditional occupation of Sweepers is sweeping out cattle barns and cleaning streets before houses, carrying away all the waste material that will go into the family's compost. In some areas, such material is considered defiling; however, no one mentioned this in Shahidpur, where there was more concentration on simple economics. The Sweeper's "special equipment" is a broom and wicker carrying basket, equipment that can be made from simple and cheap materials by almost anybody in about a half-day and which is therefore of almost no capital value.

In 1965 Sweeper was the only *jātī* name that was considered to be impolite when used in the presence of the those in the group, either for reference or address. Instead, the preferred way to speak of the group was "of Balmik" or "Balmiki"—Balmik being a great ancient poet of Sweeper *jātī*, the author of an important version of the Ramayana. By 1978, although there had been no change in the naming convention or in the sense of their relative status, Sweepers had substantially improved their living conditions in absolute terms. They had also professionalized their work by concentrating on agricultural labor and hiring out as teams for very specific actions at very specific times (hours—not just days).

In 1965 Sweeper families obtained most of their income from agricultural labor, other labor (such as construction), and from actual sweeping. Twelve families allocated the sweeping among themselves and charged those they worked for an annual fee based on a formula that reflected the amount of manure each type of animal produced daily. Farm labor was obtained by personal arrangement through the same bargaining conventions other groups used and at the same wages. In addition, one family raised pigs (actually one sow and twelve shoats at the time of the census), four Sweeper men owned musical instruments and made up a band that played at weddings, and one made a substantial income from farming (as a long-time partner to a Jat), the equivalent of the yield of 3.49 hectares.

In 1978 the musical instruments were not reported; the bagpiper and organizer of the group had died in 1977. No one was farming (Sadhu Singh, who had been the partner to the Jat, had died). There was obviously a greater need for sweeping because of the increased cattle population and two families were raising pigs (one had one; the other had twelve). But most of the Sweepers' obviously greater income came from their increased work as farm laborers.

In 1965 three Sweeper families lived in the only houses in the village that were made of unbaked bricks. These houses were rented. In 1978 they lived in brick houses which they owned, on new bricked lanes. These lanes were part of several blocks of Sweeper houses that had all been bricked over, with neat curbs, hand pumps, and areas for cattle at convenient places. Seven families reported building or remodeling at least one unit. The number of units per family increased from 1.2 to 1.56.

POPULATION DYNAMICS

The changing economic opportunities have had a direct bearing on the changing patterns of population from *jātī* to *jātī*. The connection lies in patterns of migration, which have had a far more important impact on the village population structure than patterns of birth and mortality. This can be seen clearly in Table 4.3.

Those added by birth made up 24 percent of the 1978 Jat population, 26 percent of the Harijans, 26 percent of the Sweepers, 27 percent of the Potters, and 23 percent of the Masons. These are the only groups large enough for such comparisons to be meaningful. The differences are obviously minimal. In contrast, the differences in patterns of migration are sharp, especially in the ratios of immigration to emigration. Jats and Harijans show net outflows; the other major groups all show net inflows. Most movement, as already noted, is by individuals. Only twenty-nine of the emigrants were in whole family units (six families). Of the immigrants, forty-three came in whole family units (nine families). Among Jats, the ratio of immigrants to emigrants was 1:1.87. Among Harijans it was 1:1.21. Among Sweepers it was 3:1. Among Potters it was 1.8:1; and among Masons it was 1.67:1.

A related pattern lies in the ratios of residents to nonresidents. Nonresidents are often, though not always, people away for work as short-term emigrants. The absolute number of family members reported as nonresidents increased sharply among the Jats, while it decreased among Harijans. Proportionally, between 1965 and 1978 the nonresidents among the Jat households increased from 9.4 percent of the total population to 13.5 percent. Among Harijans, the proportion decreased from 4 percent to 3 percent. Among Potters, the proportion of nonresidents increased from 2.5 percent to 15 percent.

The Green Revolution has clearly been associated with population movements, and such movements were both expected and, sometimes, feared. Many predicted that the new technologies would displace the village laborers at the bottom of the social scale, the least skilled and least educated, who

Table 4.3

Population and Sources of Demographic Change, 1965–1978 (by Jātī)

Jātī	*Population* *1965*	*Population* *1978*	*Number of births*	*Number of deaths*	*Number of immigrants*	*Number of emigrants*	*Number of nonresidents* *1965*	*Number of nonresidents* *1978*
Brahmin	32	28	7	1	5	15	8	9
Jat	367	431	104	11	39	73	35	58
Mason	43	64	15	0	15	9	1	4
Goldsmith	16	14	1	0	2	5	0	1
Barber	8	10	3	1	0	0	6	2
Water-carrier	2	2	0	0	0	0	0	1
Potter	80	115	31	0	5	1	2	17
Cotton-ginner	10	12	4	2	1	1	0	0
Sadhu	10	16	4	0	2	0	0	0
Harijan	253	319	85	11	39	47	11	9
Sweeper	49	69	18	2	6	2	1	2
Total	870	1,080	272	28	114	153	64	103

would then flock to the cities and form a politically explosive class of unemployables—as indeed appears to be happening in many countries in Latin America. In Shahidpur we see why such a disenfranchised urban proletariat has not formed in Panjab, and how it can be prevented in the future. Those who have emigrated have come mainly from the segments of the village population that were already better off and translated their relative wealth into the support of travel and, especially, education to find new opportunities when they arose. Harijans did not have as much wealth as landowners, on the average, but the reserved jobs provided a balancing advantage that led to relatively high levels of emigration for them compared to similarly poor groups without such legal advantages (like some Potter families, the Water-carrier, the Cotton-ginners, and others). Poor people without skills and education have actually flowed into the village, on balance, along with specialized artisans (Masons). And these poor people have evidently been able to raise their income levels and, at the same time, acquire skills and education. Over the long run, the village is in effect importing unskilled people, allowing them to earn incomes productively and educate their children, and then ultimately exporting skilled laborers who can, in progressive stages, move into more diverse and technical employment.

This pattern is apparently built on an older one, whereby Jats have often used external employment to gain leverage for land purchases and capitalization. Tom Kessinger's analysis of revenue documents and family histories for the village he calls Vilayatpur, in Jullundar district about 80 km northeast of Shahidpur, shows that, for the entire period between 1848 and 1968, Jats migrated out at rates that were higher than average for the village and that they made up a decreasing proportion of the village population as a whole (1974:94–96). Over most of the early part of this period, a major motivation for this emigration appears to have been the desire to earn cash to obtain more farmland in the village, in a situation of constantly increasing numbers of farms of ever-decreasing size (pp. 115–116). After 1946, however, his records show a new pattern that he describes as "the decline of the community as a referent" (p. 174). Instead of trying to find capital to bring back, the emigrants increasingly aimed to better their lives wherever their migrations might take them and began selling out their village holdings entirely. In Shahidpur this second phase evidently had begun sooner, in the emigrations to what was eventually to become Pakistan, but the overall interaction of aims and effects appears to be the same. Kessinger does not indicate that the emigration had led, by 1968, to any increase in holding size among those who remained behind, as it apparently has in Shahidpur. Perhaps this is a point of genuine difference, but more likely it is a trend that developed strongly in the region as a whole only between 1968 and 1978.

Some might wish to infer that since the present pattern of demographic changes has a long prelude, the Green Revolution is not relevant to it at all. This does not follow. The prelude rather implies that the connections between technological changes and migration are more complex than the simple "mechanization causes labor displacement" arguments at first suggest. One should also look at the ways "displacement" (migration) itself contributed to the technological changes. Past migrations and village support for the institutions they depended upon (from schools to travel agencies) contributed to the modern educational and research apparatus and the present store of capital, knowledge, and skills—ultimately creating both the necessary economic and technological base for the recent developments and a network of social relationships tying together villagers and townsmen and villagers and villagers throughout the region. The Green Revolution is thus a natural outgrowth of older patterns whereby village management groups have responded to a combination of local needs and external opportunities, and the Green Revolution in turn has both maintained these patterns and given them new shape and new significance for the future.

SUMMARY OF CHANGES

As an information system, the division of labor is less complex than the ecology but still clearly a system of high information potential. It is a message source of many elements that can be selected and combined in meaningful behavior in a very large number of ways. Its basic integrating concepts are less numerous than the distinct biological and mechanical processes that villagers formulate and attempt to respond to, although indeed the managerial concepts are subject to manipulation by social agreement in a way the biological principles are not. The 155 managerial units that each decision maker needed to recognize and select from in 1978 are a smaller number of independent elements than the seventy-odd major crop varieties, the parts of plants, the ten major animal species, the many old and new tools, and the many distinct integrative and preparatory tasks from sowing to harvesting to using each product.

Because it is more tightly integrated and especially because the integrating principles are more entirely conventional, the division of labor has less inherent stability than the ecology. There is a higher probability (and a higher awareness of the probability) that change in some single elements could entail a necessary change in every other. Since no such overall wrenching or revolutionary change actually occurred, the changes have basically followed the pattern characterizing high information systems—a bit at a time in each independent unit. Yet even in what has been said so far, we can see that the avoidance of radical alteration was not automatic but was rather the result of conscious action, such as the action that led to curbing and controlling the potential costs of the cooperative as a thrift society before it was given a more useful form by the change in central policy.

Undoubtedly one of the reasons for the different ways different laws have been received in the village lies in the fact that some laws were based on a relatively good understanding of the division of labor in such villages generally while other laws were not. Legislation would obviously be more effective if it were based on a clear and respectful understanding both of what a division of labor was in general (as distinct from an ecology, an economy, or other systems that will be described) and of the specific dispositions to be found in the communities at which the legislation was aimed. Policy which fails to make accurate assumptions about village systems of management will have its effects, if any, by a combination of accident and of the villagers' own abilities to use constructively what the policy makes available.

Prominent among the many sets of assumptions that have been used to frame legislation are those involved in the arguments for and against the "trickle down" effect of technological improvements and economic supports. Proponents argue that if one first aids the wealthier elements, their increased income will in turn benefit the poorer. Opponents argue that the effects do not actually reach the poor at all or do not reach them equitably. I too dislike the "trickle down" arguments, but for different reasons. First, they suggest that economic classes are more concrete and more separable from one another in the village division of labor than is actually the case. One class, such as landowners, could not prosper first and the others later. The system is too tightly integrated and the village families are at once too quick to grasp new opportunities and too interdependent. Second, it should be obvious that policies intended to benefit any sector or occupational group in isolation or at the expense of another, will, even if they work, necessarily produce lower overall rates of growth and more bad feeling than policies more globally and considerately conceived. What has happened in the village is not so much a trickle down as a "swirl-around," with ever-widening circles of involvement, and at the same time an increase in the total range of possibilities for virtually everybody.

5

CHAPTER

The Economy

The village economy is the system of buying and selling and of evaluating objects for purchase or sale. The Panjabi term for economics is *ārthak*. It is related to the term *ārthī*, whose main sense is "broker." Strictly speaking, the economic system is not only a village system. Villagers participate in a national system and it in turn reflects a larger, international one. The system has two main elements, each rather complex but highly organized and clearly conventional. The first consists in the system of money in use in India and in Shahidpur, which involves two basic subordinate ideas: the mathematical conventions that define the money values and the idea of legal tender and government fiat supporting it. The second consists in the market conventions by which money values are assigned to objects. These latter in turn are of two sorts. There are the conventionalized actions involved in exchanging money for a right in a good—what one says, the records one creates, and so forth. And there are the formal and standardized weights and measures appropriate to such transactions.

Economists generally define money as a quantitative medium of exchange, store and standard of value, and unit of account (see Dalton 1965). Anthropologists sometimes use similar ideas, emphasizing the idea that money is exchanged and that exchange is a social relation (see Codere 1968). These are helpful points as far as they go, but both definitions ignore the recognition of money's sanctioned conventional quality that is declared on the surface in the idea of a legal tender. Legal tender is money whose value is clearly and explicitly established by law and backed by courts. It not only can but must, by law, be offered in payment and used in records of transactions that confer rights. If a person is owed a legal debt, and the debtor offers legal tender in payment, refusal to accept such payment cancels the debt. Courts will no longer enforce it; legal agencies will no longer recognize it. The creditor is under no similar obligation to accept anything else which the debtor may offer.

The market conventions are tied directly to the status of money as legal tender. The reason the standard conventions exist and are used is that they are what the courts will recognize; the courts can only support the status of the monetary system by safeguarding the conventions. Generally the larger the amounts of money involved and the more important the goods, the more detailed and elaborate are the sale transaction and the record it creates. In a small purchase in Panjab, one usually gets a simple receipt. When a friend borrows a moderate sum of money from another, both usually go before a market official analogous to a notary and obtain a tax stamp to place in their family record books, across which the record of the transaction is written. Transfers of land must be made before a relatively high-level official who oversees the creation of an elaborate record in multiple copies.

When a person knows the form to be followed in buying or selling something, he knows how that thing is to be evaluated economically. He knows how to think of its price. In India, grain is always sold by weight, as is meat. In the United States grain is sold by volume, while meat is sold by weight. Such conventions ease or impede comparisons between goods.

When the price of one good is regularly and normally evaluated in relation to the prices of a specific set of other goods, we can speak of the several goods as being part of an economic market. This follows common usage, as when we speak in the United States of the "grain market," the "real estate market," or "the labor market." This does not follow the technical usage of economists, who require that a market be the intersection of supply and demand for a set of goods that are "identical."

Between 1965 and 1978 Shahidpur saw no change in the basic monetary system, although there were minor shifts in the legal conventions according to which economic transactions were made and there were major shifts in the patterns of goods that made up the several major markets. The main changes in the conventions for making purchases concern the greatly increased use of noncurrency instruments, such as vouchers and purchase orders, that have been associated with the increasing importance of the cooperatives. The changes in markets in turn arise partly from this, partly from the changes in available goods themselves, partly from changes in relative prices, and partly from absolute price changes as well. Panjab did not escape the worldwide inflation of the 1970s. Assigning an overall rate to the inflation is difficult and dangerous. A fourfold price increase seems to approximately cover most differences between 1965 and 1978 and to be what villagers regard as normal, but specific price changes have often varied greatly from this overall trend.

In 1965 the village economy revolved around three major markets: land, crops, and labor. Cutting across these were markets in credit, cattle, transportation, and housing. For 1978, it is necessary to add a market in aids to agricultural production other than labor and to place much greater stress on the markets for credit and housing. None of these are as they were.

THE MARKET IN CREDIT

No separate credit market was described for the village in 1965 because credit arrangements were viewed as an aspect of purchase arrangements for the specific types of goods the credit was for, hence as aspects of the several markets rather than as a single organized market in its own right. To a very great extent this was still true in 1978, but there was nevertheless a major overall shift in the relative amounts of credit available from different sources. By 1978 the cooperative system had become the main institutional source of agricultural credit and perhaps the main source of credit of any type. It was followed by personal loans from friends and relations which were still very important for nonagricultural uses, like weddings, and next by banks and merchants who offered arrangements for installment payments for such items as tractors and well sets. In 1965 the main institutional source of agricultural credit was the private *ārthī*. By 1978 they were said to be out of the business of providing credit entirely.

Ārthīs lent money against future crops but could not control what the money was to be spent on to produce them. Since the prices of inputs were uncontrolled and supplies were not guaranteed, a farmer who had borrowed money fully intending to use it to purchase productive aids might be unable to find them at the expected price—or at all. By providing most credit in kind and where possible delivering the goods to the village, the cooperative not only eliminated this trap but also reduced the costs to the farmers in both money and time. Moreover, once the state cooperatives had stockpiled goods, they were themselves vulnerable to loss and had to be sure the goods were distributed. Thus the cooperatives, and indirectly the government, inevitably assumed part of the farmers' operating risk. This led farmers in 1978 to expect a degree of responsiveness from the government that was never previously expected either of the government or of *ārthīs*.

In 1965 farmers said that *ārthīs* charged between 12 percent and 18 percent interest per year, compounded annually. (The convention for calculating 12 percent was that the borrower paid the lender one rupee every month on each hundred borrowed. If the interest was not physically delivered each month, one pice [1/100th rupee] per rupee was added to it. For 18 percent, the rate was Rs. 1.50 per month, and so on.) Among friends, the 12 percent rate was said to be most common and was the rate stated in a few account books I was shown. If a farmer sold grain to the *ārthī* and the amount received was more than the loan, the balance (or any amount) could be left with the *ārthī* and would earn 9 percent to 12 percent interest. Commercial banks were then charging just 4.5 percent interest, but required full security in collateral that could be liquidated. Various laws made it impossible to use land this way and the banks would not accept future crops, which is the reason no credit was available from them. Bank interest on savings was just 2 percent. Banks were difficult to deal with, and there were none outside the larger towns.

In June 1978, the cooperatives charged 12.5 percent interest on crop loans. In midsummer the rate dropped to 10.5 percent. The cooperatives did not pay fixed rates on deposits, but returned earned profits to shareholders at the end of each year. At the same time, according to the secretary of the Shahidpur cooperative, if one sold grain to the local marketing society, they immediately added a premium of one-half rupee per hundred to the ordinary price and if the sale amount was left on deposit with them, they paid 9 percent per annum. For long-term loans, the cooperative Land Mortgage Bank charged slightly less than the rates for crop loans, generally 9 percent or 9.5 percent. Thus credit in 1978 was not only more abundant and less risky than in 1965, it was cheaper in absolute terms.

While the cooperatives tied the market in credit very closely to the market in aids to agricultural production on what might be considered the import side of the village ledger, the state marketing cooperatives created an almost equally tight link to the grain market on the export side. One of the original ideas behind the cooperative crop loan system was that the grain would be sold back to the marketing cooperative and the loan would be written off internally as with the *ārthī*. In practice farmers still sell for cash in the marketplace that offers the best price. Outstanding loans are repaid with cash directly to their village cooperative secretary, to take to the regional cooperative office. Because of this the market for selling grain remains, institutionally, much more of a traditional free market than the market in credit, although the floor that marketing cooperatives establish is extremely important in all the farmers' calculations.

THE MARKET FOR AGRICULTURAL PRODUCE

One of the most difficult decisions a farmer has to make is how much of each crop to grow. To assess this, in principle, the farmer should look at the economic costs and gains of all possible crops in each season. In practice, however, most farmers consider just a few crops for their marketable plantings. Table 5.1 gives the basic yields and economic returns of the major market crops grown in 1965 and 1978. The lower part of the table summarizes the gains for the major rotational patterns, which is the way the farmers estimate the average annual income they can expect per unit of land.

The cost figures were given by farmers in the village. The unit sale prices for 1965 were quoted in the village and confirmed by the local papers at the time. For the 1978 sale prices, I was given some figures in the village but also told (again) that the farmers look in the paper (specifically the *Tribune*, published in Chandigarh in English) and pick the market with the highest prices, subtracting transportation costs. In this case, however, there was a discrepancy. The prices in the paper tended to be consistently 10 to 15 percent higher than those given in the village. The difference appears to be that the village prices corresponded to the announced support prices on the major commodities, the prices that would be paid by the major marketing cooperatives. The newspaper prices corresponded to the prices at which the cooperatives released the commodities, the prices to retail outlets. Farmers apparently could sell either way. Table 5.1 uses the published figures, on the assumption that they were more current. The price for rice is for unmilled paddy, not the price for milled grain which I have used previously (Leaf 1983:252). The paddy price is more appropriate; the rice is sold unmilled and on reflection the yields must have been stated accordingly, although it did not occur to me to ask at the time. The summaries show that the smallest increase for a two-crop rotation is 314 percent, for a winter crop of wheat followed by a summer crop of maize. The return from cotton followed by wheat in 1978 was 386 percent of 1965 levels and the yield from cane was 400 percent. The rice-wheat rotation was not used in 1965. Striking as these gains are, they are not the whole picture, since additional intercrops were possible between the major rotations, mainly of vegetables or potatoes. Farmers in 1978 said an intercrop of potatoes in the wheat-maize cycle could add Rs. 7,000 to the return per hectare per year, although it should be added that the potato price fluctuated greatly. In 1978/79 there was such a severe marketing collapse in the state that many farmers could not even afford to dig out the crop (Gill 1983:7).

Comparing the increases in yield with the increases in unit prices shows that the overall increases in returns do not come uniformly either from one or the other. For wheat, for example, the increased return comes more from increases in actual yield on the best land and more from increased unit prices on average land. The increased return to cotton comes mainly from increased price, but somewhat from increased yield. For maize, about two-thirds of the increase comes from increased unit prices and one-third from greater productivity. The increased return to cane is due entirely to increases in the price.

These differences reflect the different inputs each crop requires as well as the prices of the marketed material. Table 5.2 shows the components of cost for the major crops on the best land. An interested reader can compare the figures with similar reports for 1965 tabulated in my previous description (Leaf 1972:134), although two points should be noted. First, farmers in 1978 unanimously assured me that my yield figures for cane in 1965 were much too high. (There had been a boycott of the sugar mill at the time of the cane harvest, when I asked about the prices, and farmers were sensitive about being questioned closely, especially if I tried to take notes. In avoiding direct questioning I evidently took a description of yield per acre or *kilā* [3.75 *bīghās*] for a statement of yield per *bīghā*.) The figures in Table 5.2 are correct. Second, the meaning of "best" is slightly different for the two times. In 1965, unsure of the relationship between general ideas and actual performance, I tried to locate the absolute greatest yields anyone had actually obtained. In two important cases (a field of wheat grown by a very good farmer on a plot between the *abādī* and the canal and a field of maize grown by Indar Singh near his well, using hybrid seed), these were well above what one would ordinarily expect. For the 1978 figures, I was given yields that farmers could reasonably plan on for the land grades. Hence the 1978 figures are more conservative. Of course there was no way to recover comparable estimates for the earlier period.

The 1978 figures show consistently more aggressive use of fertilizers, the use of chemical weeding and insect control wherever practical, outlay for seed figured on a regular basis, and much more water consumption. The last point is perhaps especially striking, if only because anyone who has sampled the literature on the Green Revolution would anticipate the others. Wheat, for example, was stated to require three waterings in a season in 1965, against the six in 1978. Maize was allocated four in 1965 against five in 1978. Cane alone remained unchanged at ten waterings each time. Rice, in 1978, far outstripped it with the requirement of twenty-five waterings.

To decide the final cropping pattern, the returns from the possible rotations of these major crops are first considered in the light of the farmer's household needs for food, fiber, and fodder. The farmer estimates how much of the land will have to be set aside for nonmarket crops to fill out his

Table 5.1

Yields and Income from Major Crops, 1965 and 1978

Crop	*Land*	*Yield (quintals per hectare)*	*Price (Rs. per quintal)*	*Gross return*	*Gross expense*	*Net return (per hectare)*
1965						
Summer						
Cotton (*desī*)	Best	22.40	94.00	2,105.60	294.12	1,811.48
Maize	Best	33.00	94.00	3,102.00	588.24	2,513.76
	Average	20.00		1,880.00	588.24	1,291.76
Winter						
Wheat (*desī*)	Best	41.80	63.15	2,639.67	423.53	2,216.14
	Average	26.40		1,667.16	423.53	1,243.63
Annual						
Cane	Best	1,000.00	5.35	5,350.00	2,828.82	2,521.18
	Average	824.00		4,408.40	2,794.12	1,614.28
1978						
Summer						
Cotton (*desī*)	Best	24.71	350.00	8,648.50	431.76	8,216.74
	Average	20.00		7,000.00	397.76	6,602.24
Maize (*Ganga*)	Best	35.29	130.00	4,587.70	1,073.70	3,514.00
	Average	23.53		3,058.90	870.58	2,188.32
Rice (*Jāya*)	Best	74.12	100.00	7,412.00	2,276.76	5,135.24
	Average	47.06		4,706.00	2,270.47	2,435.53

Crop	*Land*	*Yield (quintals per hectare)*	*Price (Rs. per quintal)*	*Gross return*	*Gross expense*	*Net return (per hectare)*
Winter						
Wheat (HYV)	Best	47.06	130.00	6,117.80	1,158.88	4,958.92
	Average	41.18		5,353.40	889.41	4,463.99
Annual						
Cane	Best	1,000.00	13.50	13,500.00	2,794.12	10,705.88

Rotations of Major Crops on Best Land (Rs. per hectare)

	1965	1978
Wheat-Maize	4,729.90	8,472.92
Cotton-Wheat	4,027.62	13,175.66
Rice-Wheat	NG	10,094.16
Cane	2,521.18	10,705.88

Note: NG = not grown.

Table 5.2
Principal Cost Components of Major Crops, 1978 (Rs. per* Bīghā*)

Crop item	*Gain*	*Cost*	*Net*
*Maize (*Ganga *#1)*			
Grain, 3 quintals @ Rs. 130	390.00		
Fodder, 10 bundles @ Rs. 2.60 ea.	21.00		
Seed, 1.7 kg			
Waterings, 5			
@ Rs. 2 ea. (electric) or		10.00 or	
@ Rs. 6.75 (diesel)		33.75	
Fertilizer			
compost, 3–4 *rerīs* @ Rs. 10 ea.		30.00	
CAN, 42 kg		28.00	
superphosphate, 33 kg		18.50	
Hoeing, 2 @ Rs. 5 and food (Rs. 5)		20.00	
Transportation, Rs. 1 per quintal		3.00	
	411.00	Min.: 109.50	301.5
		Max.: 133.25	277.75
Rice (Jāya)			
Grain, 6.25 quintals @ Rs. 100	625.00		
Straw, 20 bundles @ Rs. 2 ea.	40.00		
Puddling			
with own oxen or			
with hired tractor		6.50 or	
Seed, 2 kg @ Rs. 2 ea.		4.00	
Transplanting		10.00	
Waterings, 25			
@ Rs. 2 ea. (electric) or		50.00 or	
@ Rs. 6.75 ea. (diesel)		168.75	
Fertilizer			
41 kg NH3		29.00	
28 kg superphosphate		12.00	
Transportation to Uncha Pind, Rs. 1 per quintal		6.30	
	665.00	Min.: 117.80	547.20
		Max.: 236.55	428.45

ecological requirements that cannot be met with market crops alone. For example, if cotton and cane are planted as market crops, some productive fodder will have to be planted on land of its own, taken out of the total. The final gain from the cane is, thus, the gain from the cane divided by the total

Crop item	*Gain*	*Cost*	*Net*
Sugar cane			
Cane, 85 quintals @ Rs. 13.50 ea.	1,147.50		
Fertilizer			
50 kg urea		80.00	
Waterings, 10			
@ Rs. 2 ea. (electric) or		20.00 or	
@ Rs. 6.75 ea. (diesel)		67.50	
Hoeing, 2–3 times, Rs. 10 ea.		30.00	
Transportation to Uncha Pind,			
Rs. 1 per quintal		85.00	
	1,147.50	Min.: 215.00	932.00
		Max.: 262.50	884.50
Wheat			
Grain, 4 quintals @ Rs. 130	520.00		
Fodder, 13 bundles @ Rs. 2 ea.	26.00		
Seed, 6.3 kg @ Rs. 2 ea. Once			
every three years.		4.00	
Fertilizer			
urea (1 quintal) or		30.00 or	
CAN, 21 kg, and		13.00	
superphosphate, 32 kg		+18.00	
Waterings, 6			
@ Rs. 2 ea. (electric) or		12.00 or	
@ Rs. 6.75 ea. (diesel)		40.50	
Weeding, chemical		10.00	
Harvest, 1/25 of yield		21.84	
Transportation to Uncha Pind,			
Rs. 1 per quintal		4.00	
	546.00	Min.: 81.84	460.16
		Max.: 103.34	438.66

land thus used for both cane and fodder. Wheat and rice, on the other hand, do produce straw which is used as fodder, so if they are planted, the land devoted to straight fodder crops can be less.

Yet taking both market and nonmarket crops into account, the total income from land held has gone up somewhat more than the percentage changes in the returns to market crops would indicate because of the increases in acreage under these crops already noted: in wheat from about 45 percent to 69 percent of winter land and of rice, maize, and cotton together from 38 percent to 51 percent of summer land. Additional gains came from the new ability to divert fodder from animals used for water lifting to those used for milk production.

Once the crop is grown, it must be sold. There have been changes here also. In 1965 farmers bagged their grain at the time of threshing. Since it was almost always sold at the marketplace in Kurali, the only decision concerned the time to sell. Later was usually better, since the price at harvest was always lowest. The grain was held until the farmer was forced to move it by storage problems or a need for cash, or until his *ārthī* advised selling.

In 1978 the bulk-storage, galvanized iron drums relieved farmers of much of the ecological pressure to sell. At the same time, the cooperative market societies had reduced the speculative spirals that one would see previously, so there was less reason to delay selling. But there were still differences in the prices paid in different regional markets. Prices in Chandigarh, especially, were said to be often higher than at the local markets. When a farmer had time and was ready, he bagged what he wanted to sell, took it by cart or perhaps hired truck, and sold it wherever he found the greatest return over transport costs.

When agricultural products are used to pay debts in the village, they are evaluated at the current market prices, although in the nature of the case there is always some room for discussion about just what those prices are.

THE MARKET IN LAND

Land is the most important good regularly bought and sold in the village. As previously, in 1978 there were basically five methods for obtaining it. These were birth, purchase, mortgage, tenancy, or holding it on trust (someone else's land held in one's own care). The methods represent decreasing levels of cost but also decreasing degrees of security in the order listed. Purchase is the most expensive way to obtain land. Tenancy is much cheaper. Holding land in care involves no money outlay. Since 1964 both the costs and the relative frequencies with which each arrangement is used have changed. Costs of ownership have gone up slightly out of proportion with tenancy and out of proportion with the increases in yield, but much more land is farmed by its direct legal owners and less on other arrangements. Tenancy has become very uncommon.

In 1964/65 farmers quoted the sale price of the best grade of land at about

Rs. 5,000 an acre, Rs. 17,650 a hectare. In 1978 the same farmers quoted the price for the same grade at Rs. 76,000 a hectare. I had checked the 1965 reported price against previous actual sales and found it consistent (Leaf 1972:117). Although I had no opportunity to make a similar check for the 1978 figure, I am sure it is equally reliable. The price for well-irrigated land in 1965 was given at the equivalent of Rs. 9,400 a hectare; in 1978 it was put at Rs. 47,000. There was no 1978 quotation for the poorest (rainfall) land. Villagers said there wasn't enough of it sold to form a pattern.

These increases of 430 percent in the price of the best land and an even 500 percent for average land may seem astonishing (if one isn't aware of land prices in most of the United States and Europe over the same period), yet in fact they very closely follow a pattern previously established. The inflation of land prices corresponds very closely to interest rates. If a farmer owning either grade of land had sold it in 1965 and lent the money out at the prevailing interest rate of between 12 and 14 percent compounded annually, by 1978 his principal and accumulated interest would have been very nearly equal to the market prices of the land. In 1965 the same thing had been true of land prices over the preceding ten years, both as attested by farmers and as shown by the official records (Leaf 1972:123).

Economists in the past have often looked at the high rural interest rates in India and similar countries and at the low "economic" profit margins (based on their own assignments of values) and argued that the farmers must be economically irrational. If they were not, the argument goes, they would realize that they would be better off selling their land and lending out the money on interest. In reality the situation is a little more complex. A farmer who sold out would have had the interest and principal, to be sure, but the farmer who kept his land would have had the same gain in the appreciated value of the land. Moreover, the farmer who sold out would have had to take his living expenses out of the interest returns, while the farmer holding the land would have taken his living expenses out of operating costs *before* the gains in appreciation of land value were realized, as part of his overhead, and he would probably have had profits besides. Thus he would actually have more. The economists' calculations pretend that personal and family maintenance are irrelevant to the problem; the farmer knows they are not. Selling out only makes sense if someone has something substantially more profitable than farming to invest in, or more profitable than farm work to devote time to while his money is earning interest.

In addition to being closely tied to interest rates, land prices have also been closely related to the value of what the land can produce (again despite many economists' declarations to the contrary). In 1965 farmers explained the increasing value of land by reference to the market value of the most remunerative crop, cane. It was considered possible to buy a small amount of the best land and expect to earn back its cost in three to four years—not counting the expected appreciation of the land price itself (Leaf 1972:118). Of course, it was necessary for a farmer to already have a going operation, especially other land that could be used to take care of his family's daily needs and all the tools required not only to farm the old land but the new as well. Farmers assumed, in quite a direct sense, that the purchase of land was a marginal cost for capital—an incremental outlay to make an incremental improvement in one's basic operation. The calculation was that marginal return would exceed this marginal cost in three to four years.

In 1978 the relation between marginal expenditure and marginal return was only slightly less favorable. Assume a farmer bought a *bīghā* of average land in 1978 (a hectare would be much too much for this reasoning to apply). The price would be Rs. 3,500. The most profitable rotation was cotton-wheat, earning a total return of Rs. 943 per year (including the value of the straw of the wheat). At this rate, the value of the land would be earned back in 3.7 years. With a wheat-rice rotation, net earnings (including the value of straw) would be about Rs. 648 a year, which would repay the price in 5.4 years. For the least profitable rotation, maize-wheat (which could, however, allow an intercrop of potatoes), the earnings would be Rs. 554 a year and the repayment would take 6 years. By Western standards, and by Indian urban or industrial standards, these are not at all bad rates of recovery. Price increases for the commodities and inflation of land value itself would reduce the time needed for repayment.

Looked at in this way, it is almost surprising that land is so cheap; however, two final points must immediately be made. First, few farmers have their needs so well in hand that all they require is more land in order to increase production. Second, in 1978 farmers said that the land prices had been forced up most recently not by farmers themselves but by Harijans who bought very small parcels for vegetable gardens and for dairying, both of which are more remunerative than ordinary farming operations.

Land sales are formalized before a Tehsildar or Naib (vice-) Tehsildar. The Tehsildar is the civil service officer responsible for the central administration and revenue of a Tehsil and is a Panjab Civil Service officer of considerable importance. There are two Tehsils in Ropar district and Ropar is one of seven districts in the (post-1967) state of Panjab. Both buyer and seller must appear in person at the Tehsildar's office with witnesses and with the village revenue clerk. The witnesses have to be known to the officer and/or be officially designated witnesses of record for the village, such as panches. The witnesses attest that the buyer and seller are who they say they are and that the price of the land is its true value. All parties have to agree publicly to the identity and description of the piece or pieces of land as identified in the records maintained by the revenue clerk. The buyer and seller must both affirm that they agree to the transaction freely and are under no coercion of any kind. If there is anyone with a right of preemption who can stand between buyer and seller, they should be present to agree to the transaction. Only if the officer is satisfied on all these points does he let the transaction proceed. There are three main records involved: the record of rights, the mutation record, and the genealogical record. The record of rights is the general compilation of all landholdings in the village and is in principle updated for the whole village every eight years. Transactions which occur between major rewrites are indicated in marginal notes referring to other records, and changes in ownership are sketched in with red ink. There are three copies. One is in the Tehsildar's office; one is in the district office; and one is the village revenue clerk's copy. The Tehsildar's copy and the village copy will be changed; the district copy will be changed later upon receipt of the appropriate notifications.

The marginal notes in the record of rights name the type of transaction and refer to a transaction number in the mutation record. This is a book of forms with places to note most of what is attested: the description of the land, the buyer and seller, the price, and the location of the entry in the record of rights. When everything has been attested and the money physically counted out and handed over before the official, each party to the transaction signs his copy of the page in the mutation record across a dotted line marking the border between the receipt and the stub. The copy is then torn off along the line so that the signature is split lengthwise. If there is any dispute, the parties will have to bring their copies back and match the upper and lower halves of the signature. Finally, the genealogical record will be changed (in red until a major redrawing) to conform to the record of rights. All of this makes it highly unlikely that these most important records will be falsified.

The arrangement called mortgage (translating the term *rahin*, literally "remaining") is different from what the term would suggest in English. In it, the landowner borrows money using land as security. The amount borrowed is exactly one-half the sale price of the land. In exchange, the convention is that the owner gives up the use of the land instead of paying interest. The lender farms the land and keeps the crops. He may, however, rent it out—even rent it back to the owner. When the landowner wishes to redeem the land, he returns the amount borrowed and reclaims possession after the current crops are harvested by the lender. If he decides not to redeem the land, the lender has first option to buy it and he pays only the balance of the agreed-upon sale price—the amount lent once again. Because of the tight link between loan and possible purchase, the valuation of the land is very important and has to be agreed upon at the outset. The transaction is conducted before the Tehsildar in the same manner as a sale, except that no entry is made in the genealogical record.

Tenancy is based directly on the productivity of land and only indirectly on the land price. The linkage to productivity lies in the fact that the baseline model of tenancy is sharecropping. In 1965 this sharecropping was stereotypically described as a relationship between a landowner and a landless tenant. The formula was that the landowner provided the land and wells, the tenant provided the draft animals and all the labor. Farmer and tenant divided the expenses evenly and each took one-half of the crop. This meant one-half the portion that was marketable, not fodder, because (again stereotypically) the "landowner has no cattle." This arrangement is called *batāī*, more specifically one-half *batāī*. A variant form of *batāī* was *panj-do-banjī*, meaning "two out of five arrangement." In this, the tenant put in three-fifths of the expenses and took out three-fifths of the gain; the owner put in two and took out two. *Panj-do-banjī* was used for land growing cane, because of the crop's greater labor and draft power requirements. By 1978 the increasing water and labor needs of the new crops had eliminated the differential and all sharecropping was on one-half *batāī*.

Cash rents in 1965, following a pattern well attested throughout India, seemed at first to be very low when calculated as a percentage of land prices—running between 2.7 percent for the best land and 4.1 percent per annum for rainfall land. Yet the rents were described by farmers as based strictly on *batāī*. The cash rent for a *bīghā* of land was equal to one-half the probable market value of the marketable portion of its yield, less about 20 percent. At that time, a *bīghā* of average land would produce about Rs. 100 to Rs. 110 worth of wheat, so the owner's share with *batāī* would be worth Rs. 50–55. The cash rent for the same piece of land would be about Rs. 40 a year. The reason cash rent is less than rent in kind is that the cash rent is paid at the beginning of every year and therefore is paid even if the crop fails. The share rent is paid at the time of the harvest and thus cannot be paid if there is no crop. One year in five was considered a loss year, when the share in kind would not be collected but rent would be. Hence in the long run the rents were equal. It should be noted in this that there is no mention of a second market crop. Since I had expected the formula given to be only a rule of thumb, to be implemented in different ways in different contexts, and since I was mainly interested in establishing the bare existence of such rules and the logical connections among them (none of which was recognized in the literature at the time I began the study), I did not question *how* they were used in detail at the time and in 1978 it was too late to do so. But the evident reason relates to the general fact that most land in 1965 was, out of necessity, devoted to nonmarket crops. Thus for most land, the one crop of wheat in the example (which I have related exactly as it was first told to me) was the probable annual pattern.

In 1978 cash rents were slightly lower as a percentage of total land price than in 1965. The rate for the best land was about 2 percent of sale value, that of average land was about 2.6 percent. It was Rs. 125 a *bīghā* for the best land (equal to Rs. 1,483 a hectare) and Rs. 100–110 (Rs. 1,185.60–1,305 a hectare) for average land.

When I was first given these rates in 1978, I naturally calculated the value of *batāī* in order see if the old relationship still held and if it justified the abandonment of the differential between half-*batāī* and *panj-do-banjī*. It did not work. The value of the share of *batāī* retained by the tenant was substantially more than the cash rent he would have paid. I took this result back to Indar Singh and the others who had given me the figures. Had the system broken down? Had the relationships become irrational? No, the system was exactly the same, as I had been told. Cash rent was still based on *batāī*. Only now the tenant paid the taxes and *all* the seasonal expenses. I naturally tried to probe into why the arrangement had changed in this way, and the response showed that the relationship between owner and tenant had been entirely reconceptualized. My question was whether the well was included in the land (which was true for the earlier period). The answer was no, because the tenant "has his own." That is, instead of being represented as a relation between landowner and nonowner, the *batāī* relationship was now conceptualized as a relation between two landholders. Again, this is nothing other than a convenient rule of thumb, like a standard lease form one buys in a business supply store and modifies as needed. The implication of this particular form is that if the tenant does not have a well, he will have to buy water—probably from the owner. Then, of course, his expenses might be higher still.

The cash value of *batāī* rents for 1978 can be calculated from Table 5.2. For wheat on the best land, for example, the tenant will keep one-half the value of the grain and all the straw, and will pay all the expenses listed plus taxes (Rs. 27 a hectare). The result will be a return to the tenant of about Rs. 1,712 per hectare. The stated cash rent is about 87 percent of this—which is evidently a very precise equivalent of the old standard, conceptually, since risk must have been reduced.

The constraints on the interpretation of these general formulas can be indicated by comparing these wheat-based calculations with those for the most-remunerative rotations (these were *not* offered by farmers as examples in either period). In this case cash rent is actually just 12 percent of the net gain from the maize-wheat rotation, and 9 percent of the net gain from the rice-wheat rotation on the best land. Why would a farmer not rent land, pay cash, and grow these rotations? The main answer is that no one will rent the land to him—as will be seen shortly. But leaving this aside for the moment, if the tenant did succeed in obtaining land, he would have to supply his own draft and human labor needs and to do this, he would have to set some of the land aside for nonmarket crops just as a farmer does who is working his own land. In the end, because of the ecological constraints on each family (which limit the ability to grow for market especially sharply on smaller holdings), the proportions of market to nonmarket crops on rented land would have to be about the same as on land that is cultivated by its owners—so that, for example, wheat would be grown on only about 69 percent of the winter land and the other market crops on about 51 percent of the summer land.

The calculation "as if" wheat were the only crop divided in *batāī* is obviously a convenient surrogate for stating all the usual portions of wheat *and* the usual portions of other crops (if one calculates it out, the relationship is quite close). Any arrangement will take it as a baseline and any radical departure from the equivalent of all wheat in the winter and all nonmarket crops in the summer will require additional discussion and special mutual agreement. For example, a tenant with buffaloes might want to grow all fodder. On a cash rental calculated as if the crop were wheat, the owner would have little cause for concern. But on *batāī* the owner would obviously not agree unless he too had cattle and needed the amount of fodder that the tenant would provide.

Tenancy arrangements are described (very briefly) in the record of rights and are entered into before the Tehsildar. The amounts of income obtained by village families on the basis of each type of arrangement in 1965 and 1978 are given in Table 5.3. The arrangements are: single ownership and joint ownership (combined in this table), tenancy (including tenant of owner and tenant of mortgage holder), partner of owner, partner of mortgage, partner of tenant, tenant of tenant, land held "in care of" for a tenant, and others. The reason for combining the two types of ownership is that the costs and gains of the arrangement are the same. The same reason applies to the several types of tenancy at will. On the other hand, the several types of partnership all reflect different inputs by the partner and different shares of the total crop, as does tenant of tenant and the "in care of" relationship.

Building on the local definitions in order to avoid the difficulties economists have regularly gotten into by stating income in monetary terms, I have calculated agricultural income by stating it as a portion of the area held. That is, for example, a full owner farming his own two hectares is credited with two hectares of income. A mortgage holder on two hectares who farms the land and takes all the crops will also be credited with two hectares of income. If the owner rents the land on *batāī*, he would be credited with one hectare and the tenant with the other, even if the rent was paid in cash. If the

Table 5.3

Sources of Agricultural Income by Landholding Arrangement, 1956 and 1974

	Ownership	*Mortgage*	*Tenancy*	*Partner of owner*	*Partner of mortgage*	*Partner of tenant*	*Tenant of tenant*	*"In care of" land for tenant*	*Other*	*Total*
1956/57										
Proportion of all land	0.57	0.09	0.31	0.02	0.00	0.01	0.00	0.00	0.00	1.00
Number of families	96.00	28.00	74.00	5.00	3.00	5.00	1.00	0.00	0.00	212.00
1974/75										
Proportion of all land	0.88	0.03	0.05	0.00	0.00	0.00	0.00	0.01	0.02	1.00
Number of families	115.00	8.00	19.00	0.00	0.00	0.00	0.00	5.00	6.00	153.00

mortgage holder would happen to rent the land back to the owner, each would similarly be credited with one hectare.

The table summarizes the official records of rights that were current in the village in 1965 and 1978. The actual years in which these records were drawn up were 1954/55 and 1974/75 and it is to these years that the details most clearly apply. (I ignored the interim notations for these tabulations because they do not provide a complete description of changes without the supplementary records.) The rightsholders are all those listed in the records, both resident and nonresident.

The most striking change is that in 1965 just 57 percent of agricultural income came from land held on ownership, while by 1975 such land provided 88 percent of all farmers' income. The rest came from land held on mortgage, tenancy, partnership, or some combination of these. There are four basic reasons for this change. Just one is economic in the strict sense. One is historical. Another is managerial and the fourth is ecological. Historically, the contested consolidation was completed. Thus the personal interchanges of land that farmers had often made to bring their holdings into convenient units were no longer necessary. The managerial reason is that changes in legislation intended to advance land reform were perceived by farmers as greatly increasing the risk that a tenant might obtain a legal claim to be a "hereditary tenant" on the land he rented. If this happened, the owner could neither reclaim the land for his own use nor raise the rent in accordance with increasing prices or market conditions. Economically, increased capitalization removed the need for the policy of making the maximum possible use of one's own family labor in order to minimize the cost of hired labor. At the time of the first study this policy had persuaded relatively large families to take larger amounts of land on tenancy or mortgage rather than buy much smaller amounts. In this way they avoided, in effect, wasting their human resources. In 1978 hired labor was better organized and more regularly available, while there were many other profitable ways to employ the members of the farmer's own family. The ecological reason was elimination of the need for short-term use of rainfall land to grow fodder.

Expensive as agricultural land was, it was much cheaper than land for house sites in the *abādī*, which was quoted in 1978 at Rs. 20,000 a *bīghā*, or Rs. 235,295 a hectare. This, together with considerations of comfort, was a main reason why so much new building was being dispersed into the fields.

As in 1965, the market in land in 1978 was extremely well ordered and important, closely tied to the interest rate and to productivity. Despite changes in price and some strategies of acquisition and use, land continued to be one of the most important tangible forms of both investment and savings.

THE MARKET IN CAPITAL IMPROVEMENTS

The market in land is closely related to the market in means for improving its productivity or profitability. Farmers wanting to increase production must compare the advantages of renting some additional land with those of buying a well pump, or compare the benefits and costs of selling some well land in order to buy canal land with keeping the well land and adding to pumping capacity.

In describing the situation in 1965 and the prospects for improvement as they appeared then, I summarized many of the responses of farmers by saying that any new development program would have to provide a way to relieve the obvious energy shortage then in existence: for increased water lifting and distribution, for transport for all kinds of inputs, for new production and processing of crops in the fields (Leaf 1972:255–258). Immediately upon division of the state into Panjab and Haryana in 1967, there was a widespread removal of licensing restrictions on local light manufacturing industries that had just this effect by producing the range of new mechanical devices that has been noted. Since most of these devices are powered by pump motors, the favorable price of these motors has been a major key to the process of mechanization.

The most popular 5 horsepower diesel engine cost Rs. 3,000 in 1978. It required one liter of fuel for one hour of operation, which would water about 0.7 hectares. Diesel fuel was sold at a government regulated price of Rs. 1.40 a liter (about a third the price of gasoline). A 5 HP electric motor cost the same as the diesel. The additional fittings and pipe to install a complete well would bring the cost to Rs. 7,000. The cost for electricity in 1978, Rs. 65 a month, was equal to 46.5 hours worth of diesel fuel. This is the economic aspect of the water supply situation that made rice and vegetable growing much more prevalent among owners of electric motors than among owners of diesel pumps.

Apart from the motor-operated machinery, the major tools in the village were the four tractors and their attachments noted in Chapter 3. One 50 HP tractor was owned by Mehima Singh, an uncle of Naranjan and a former member of the panchayat. Since it is widely held in scholarly and administrative circles that it is uneconomical to own a tractor of this size (or just about any size) with holdings as small as those in Shahidpur and since I knew him to be a good farmer, I was curious to hear his own calculations.

Mehima explained that he bought the tractor, a year previously, not in spite of having about three hectares of land but because he did. The land did not yield enough income, but since land in the village was scarce, he could not afford to buy more. The tractor was in his view a better investment. The new tractor allowed his own farming operations to be completed so rapidly that he could take advantage of the busy sowing season and hire it out for custom work, with his son as the driver. Mehima estimated that the tractor could plow an acre in fifteen minutes (his exact statement; this is equal to one hectare in thirty-seven minutes). Oxen, by contrast, require a full working day to plow about 2 acres, or 0.8 hectares. The rate he charged for plowing was Rs. 25 an acre (Rs. 62 a hectare), which could yield Rs. 800 in a day. For cultivating, he charged Rs. 30 an acre. These rates are comparable to, or slightly less than, rates for the same operations with oxen, despite the saving in time to the customer. Mehima reported that he had plenty of business, though not in Shahidpur where the holdings were too small.

If one expects only three months of work a year and only twenty working days a month, the annual return to a tractor hired out this way would still be Rs. 50,000. This is the equivalent of the annual gross income from about 3.5 hectares of land, although at from Rs. 40,000 to Rs. 85,000 (including plow, cultivator, and equipment for puddling rice fields) a new tractor costs no more than the equivalent of one and a half hectares. In Mehima Singh's case the tractor was bought used, for a cash price of about Rs. 42,000 total.

The tractor is obviously far from a purchase made for "show," or "status," as some observers have suggested. It is a good investment and would remain one if it worked for only four hours a day, twenty days a month, three months a year. It is a good investment because: (1) the farmer's family labor, with the help of the tractor, is not needed on his own land; and (2) others do not have tractors.

Two of the other tractors, one large and one small, are owned by families in similar circumstances. One owns just under one hectare and farms about two-thirds hectare. The other owns and farms just over 3.5 hectares. Only one tractor (a brand new 47 HP International) belongs to someone with "enough" land for it by official standards—the Mahant.

Chemical spray pumps are now common and their prices are amortized in the cost figures for crop production given in Tables 5.1 and 5.2, as are the fertilizer, insecticide, and herbicide costs themselves. None of these were used in the village in 1965.

In 1978 the storage drums used for grain cost Rs. 325 each and the government would issue a permit for a 50 percent subsidy. Since the ten quintals of grain they held were worth anywhere from Rs. 900 to Rs. 1,300, it was very easy to imagine that they could pay for themselves in just a few years, even without the subsidy, by letting farmers take full advantage of annual price fluctuations, even though the cooperative buying had for the moment stabilized grain prices to the point where the annual variations were about 10 percent, rather than the 10 to 50 percent that regularly occurred around 1965. (In 1980, however, a combination of bad weather at harvest and a blight greatly reduced the wheat harvest and once again set in motion major price increases.)

THE MARKET IN LABOR

Some of those who claim that the Green Revolution has only made the rich richer and the poor poorer have argued that the wages of agricultural laborers have not increased as much as the income to farming operations. In one aspect, this is true: the actual hourly rates have increased only modestly. But annual income is the hourly rate multiplied by the total hours worked, and this has increased dramatically.

In 1978 as in 1965, labor was hired both directly and implicitly. For example, the labor of Sweeper families was hired implicitly when villagers contracted with them for the service of cleaning out barns and maintaining compost heaps. Potter labor was hired implicitly when a householder contracted with them to supply clay vessels of a set number and type for a year. Direct hiring was done on a daily or yearlong basis for the performance of some specified task or tasks, rather than for the delivery of a specific product. Both types of hiring were called *theka*, translated as "contract." There were basically two rates for labor hired directly. Village artisans, mainly the Masons, earned one rate at various tasks that included carpentry, tool repair, running milling equipment, running threshing equipment, and construction. Village laborers earned the other, lower, rate at various tasks that included field operations like weeding, plowing, threshing, sowing, and cutting cane, as well as nonagricultural work such as well digging, canal and road repair, and acting as hod-carriers and general assistants in construction work. Since laborers and artisans both regularly worked outside the village as well as in it, local rates followed rates in nearby villages and towns.

Pay for direct labor always involved two components, cash and food. In 1965 the cash rate for artisans was about Rs. 5 a day. The cash rate for laborers (stereotypically spoken of as Harijans, but not necessarily Harijans in fact) was Rs. 2 a day. The food component was the same for both: three meals and two teas for a full day and two meals and one tea for a half-day.

To arrange for most types of field labor, the farmer met with the prospective workers in advance of the appointed day, agreed on the work to be done (after an expected argument), and paid the cash portion of the wage in advance. For harvesting, the entire payment was made in kind and less food was given, just one meal and two teas. The rate of payment depended on the quality of the crop. It was one bundle in seventeen for poor crops, one in twenty-five for the best crops. As they cut the crops, workers tied the bundles using ropes of a fixed length supplied by the farmer. The bundles to be given in payment were selected by the farmer at the end of the day. The reason for the different ratios was that the better crops grew more densely, so one could cut more in a day. Compensating for this, the rate was intended to yield one bundle for a *bighā* and a *bīghā* was considered to be the amount of land a single person could clear in a day. The smaller amount of food was to counterbalance the fact that cash value of a bundle was much greater than the usual Rs. 2. The farmers realized that they had to pay more in times of extreme demand, but tried to come as close to a parity with the ordinary labor rates as possible.

In 1978 the cash component of the rate for skilled labor was stated as Rs. 20 to Rs. 25 a day. The cash component of the rate for field labor was given as Rs. 5 a day. Food was two meals and one tea for a full day. For harvest, the negotiation and advance visit to the field were eliminated. Crews charged a flat rate of one bundle in twenty-five. The harvest rate is still the equivalent of one bundle a *bīghā*, but the harvest season is extended.

From the laborer's point of view, the main gain has come not from the increased labor rate, but from the increase in days of employment per year. The higher levels of water application and fertilization created a need for more weeding and maintenance work and there was much more to harvest. Additional crops required more clearing and plowing operations and the new construction of houses, wells, and roads created competing demands for the same labor pool. There were also some entirely new types of work, among the most important of which was transplanting rice.

The absence of records and the nature of the work make it very difficult to survey the actual income of a large number of laborers. However there is a ready alternative. In 1978 as in 1965, to avoid hiring the relatively expensive day laborers, farmers hired yearround contract workers whenever they could. Presumably, in such an arrangement the worker would not take less than could be expected from day labor on a yearly basis, so his minimal gain would be security of employment while the farmer would obtain more work days for his money. In 1965 the rate for a person of minimal experience, such as a boy to tend buffaloes, was all meals every day, two suits of clothes, two pair of shoes, and Rs. 100 cash. One farmer once estimated that an ordinary laborer would not earn more than Rs. 300 to Rs. 400 a year. In 1978 the rate for the same sort of person of minimal experience was two suits of clothes, two pair of shoes, medical expenses, and Rs. 1,000. For a more experienced worker, the cash component could go as high as Rs. 2,100. In 1965, assuming a person ate the equivalent of a kilo of grain a day (this is a generous estimate), the value of the food component would be about Rs. 334. Adding this to the clothes and shoes would bring the total income cash equivalent for 1965 to about Rs. 490. The comparable figure for the lowest-paid worker in 1978, leaving aside medical expenses, would be about Rs. 1,620. In an economic sense, this is obviously why laborers are coming into the village rather than leaving it. It also explains the farmers' view that laborers received a relatively greater share of total production in 1978 than in 1965.

THE MARKET IN HOUSING

As already suggested, the market in housing is also a market in places of work. Village buildings are capital at least as much as consumption and expenses for them are equally chargeable to overhead as to profit.

In my census of 1965, one of the questions about house ownership was "Is this house the ancestral property of the *owner*?" In almost every case, the answer was yes. In absolutely every case, the name given for the *mālik* was the same as for the owner. The obvious interpretation was that housing was seldom if ever bought or sold. In 1978 I replaced these sections of the form with questions about changes in housing since the first census. What quickly emerged was the pattern of active growth and exchange already indicated. Of the 162 families that owned at least one unit of housing, 72 had had at least some new housing construction since 1965. This ranged from putting a concrete floor in a single unit to the construction of entire, new, multi-room buildings.

The best estimate of the monetary value of housing is obtained from the cost of current construction. There are no official records of housing titles, leases, and the like comparable to the records for agricultural land. At the same time, a personal survey would be distorted in the way that the first census was. Owning the house one lives in is apparently part of being "master" of the house, in a stereotypic sense, and is a matter of some pride. A further point is that many of the transfers were among relatives and most of these were rather complex exchanges in which money played only a limited role.

For construction, the general procedure is to contact a Mason and describe the project. The Mason himself figures the amount of material that will be used for walls, ceiling, floors, windows, and doors and estimates the cost on the basis of the price of these materials and the cost of his own time and a crew of assistants, generally three assistants for each Mason working. There is a standard expectation that the Mason will lay a thousand bricks a day, and the relationship is fairly fixed between the number of bricks and other components of the construction, which amounts to a set understanding of both the cost and rate of construction per square meter of wall, floor, and ceiling. The material calculations were the same in 1978 as in 1965, but the prices had changed in two ways: they had increased and government price controls had created side-by-side legal and "black" markets. Table 5.4 gives the prices of the main items for 1965 and 1978.

The increases add up to considerably more than the fivefold increase in the price of land, but again this does not tell the whole story. The quality of construction is now higher, which means specifically that construction now regularly includes full concrete floors and concrete plaster on the interior (and sometime exterior) of the brick walls. In 1965 usually only ceilings and the floors of storerooms were cemented this way. The cost of a concrete floor is about the same as that of a wall. Windows that formerly had only bars now commonly also have frames and screens and even sometimes glass jalousies, an extra expense. In 1965 I roughly estimated on the basis of current building costs that the value of the village building stock was about 90 percent of the total value of all village land. If the old buildings are evaluated at the replacement cost indicated by the new materials and an allowance at the same rate is made for the new buildings, the total should come to more than the current value of all land. The buildings would never be sold all at once, so the figure is of no practical significance with respect to possible decisions that will be made in the village, but it does indicate the very great importance of this market.

In 1978 the housing market was somewhat isolated from the other major markets because no credit was available either for construction or purchase. An official of the Land Mortgage Bank office in Chandigarh explained that the reason for the cooperative policy was that expenses on houses were deemed to be for "nonproductive purposes." I do not know why credit was not available from commercial sources, but I suspect that besides being related to a general lack of imagination on the part of lenders, it has something to do with the difficulty of fixing a viable claim on ancestral property in case of default. Thus despite the increased costs and the great total flow of money involved, up to 1978 all dealing was strictly for cash.

THE MARKET IN CATTLE

In 1965 cattle were generally bought for cash either locally or at nearby fairs, at which buyers and sellers came together from different regions and transacted their business in the presence of market officials who issued the appropriate stamps and records. In 1978 the system was basically the same, but farmers commonly traveled much farther to get better animals and cooperatives were involved in the market both on the supply side and in the provision of credit on the purchaser's side. The costs had also changed considerably, both absolutely and relatively.

Table 5.4
Costs of Building Components, 1965 and 1978

Building components	*1965*	*1978 (controlled price)*	*1978 (uncontrolled price)*
Labor			
skilled (bricklayer)	Rs. 5, 2 meals, 2 teas		Rs. 20, 3 meals, 3 teas
unskilled (hodcarrier)	Rs. 2, 2 meals, 2 teas		3 laborers, Rs. 5 per day ea., plus 3 meals, 3 teas
Bricks (per thousand)	Rs. 15	Rs. 105, plus Rs. 20 for transportation	Rs. 120, plus Rs. 20 for transportation
Cement (per quintal)	Rs. 17	Rs. 44	Rs. 80
Iron beam (per foot)	Rs. 4.50	Rs. 16–17	Rs. 16–17

Note: Mason lays approximately 1,000 bricks per day.

The purchase prices for the major types of cattle for 1965 and 1978 are given in Table 5.5.

The shifting prices reflect shifting uses as well as the general inflation. For operating the Persian wheel, the camel was the preferred animal in 1965 because it could work blindfolded. Since it could not see that there was no one directly behind it, farmers were able to work nearby and prod the animal simply by making noise. Bullocks, by contrast, would not tolerate a blindfold and if they were not followed constantly, they would stop. For all other tasks, however, the bullock was easier to use. In 1978 draft power was no longer needed for lifting water, but it was still used for plowing, leveling, sowing, fertilizing, and cultivating—and even occasionally for threshing. Accordingly, the price of the camel had actually dropped in absolute terms, but even at the new low price there were no buyers in the village. The bullock price, by contrast, had increased, though not so much as the price of land or housing.

The goat prices increased relatively more than the bullock prices and directly reflect the increased price of meat from Rs. 2.50 a kilo to Rs. 12. In 1965 the goat prices mainly reflected their milk production; thus the price change is a major indication that there have been qualitative gains in welfare. The pig prices are also based on their value as meat, as the table notes directly.

The change in the types of "cow" represents another new development. The "American" type is a hybrid, the first type of cow that has been able to compete with the buffalo as a milk producer. It is priced accordingly. In 1978 there were only two in the village (one unreported in the census)—one owned by a Harijan family, the other tethered near the house of Banarsi Das.

The major milk animal, as already noted, continues to be the buffalo. It is difficult to determine why its price has not risen proportionally with that of land or houses, unless it is simply that buffaloes are in relatively greater supply. In 1978 they were often purchased by cooperative loans, as were hybrid cows, as part of "dairy units" in a large scheme organized under Milkfed. The program, which began in the area in 1972, also includes a regular system of milk collection and loan repayment. Milk is picked up in the village by truck each morning and evening to be taken to a chilling plant in Morinda. Farmers receive spot receipts as the milk is weighed. Every eight days a clerk comes by and will give either cash for the full value or credit up to half of the value against the farmer's loan. The price was Rs. 2.50 a kilo, which is to say perhaps Rs. 13 a day on the average per animal, or about Rs. 4,000 a year. This very substantial sum is precisely the reason several farmers shifted from agriculture to dairying. In 1978 one farmer owned eight buffaloes, seven farmers had four each, and seventeen families (not all Jat

Table 5.5
Economic Characteristics of Cattle, 1965 and 1978

Cattle	*Life span*	*1965 prime price*	*1978 prices*	*Annual fodder consumption*	*Yield 1965–1978*
Buffalo	6–10 yrs.	Rs. 800	Rs. 2,000–3,000	54.0 quintals	gives milk at rate of 10 kg per day max., 1 kg per day min. at year end[a]
Bullock	10–12 yrs.	Rs. 500	Rs. 1,500–2,500	27.0 quintals	one pair can pull 9–12 quintals in a *gaddā*, or 20 quintals in a *rerī*
Camel	12–15 yrs.	Rs. 1,000	Rs. 500	54.0 quintals	can carry approx. 4 quintals or replace two bullocks lifting water
Donkey	12 yrs.	Rs. 200	———	13.5 quintals	carries 0.75 quintals per pack
Goat (doe)	NR	Rs. 30–60	Rs.200–350	13.5 quintals	milk 0.5–1.0 kg per day, then meat
Mule	20 yrs.	Rs. 3,000	———	27.0 quintals	carries 1.5 quintals per pack
Pig	NR	——	Rs. 8–14 per kilogram	——	meat, bristles
Ram	NR	——	Rs. 200–350	13.5 quintals	meat @ Rs. 2.50 per kilogram in 1965; @ Rs. 12.00 per kilogram in 1978

Note: NR = not relevant.
[a]Less in 1965; see text.

by *jātī*) had three each. (Three is one dairy unit for cooperative purposes.) In 1965 the largest number anyone had was six. Three families had five each, three had four, and seven families had three. The rest were owned singly or in pairs.

THE MARKET IN TRANSPORTATION

In 1965 the cost of transportation to and from nearby market towns limited farmers to selling goods that were compact and that had high values in relation to weight. By 1978 the cost of transportation had effectively decreased. Table 5.6 gives the basic costs for the major means of commercial transport, oxcarts.

In some parts of north India, the terms *gaddā* and *rerī* appear to be interchangeable. In Panjab they designate two different types of vehicle. The *gaddā* is the traditional oxcart with large, narrow, wooden wheels and a high, wooden axle, sometimes with the hubs sheathed in metal for longer wear. The *rerī* is the newer type, built on an automobile or truck rear axle and tires. The two types of *rerī* have different-sized wheels. Farmers are willing to pay more for the lower-wheeled type because it is easier to load and unload. All the oxcarts require the same draft power, one pair.

The table shows that the purchase price of *rerīs* is in absolute terms two and one-half to three times what the price of *gaddās* used to be. However the capacity of a *rerī* is itself over twice that of a *gaddā* and a *rerī* lasts more than twice as long, so the price per unit of capacity per year over the long run is actually about half that of a *gaddā* in absolute terms.

The rental rates show a similar pattern. The day rate per unit hauled in 1978 was just over twice that of 1965. But here again, one should note that two and one-half times as much could be carried in a single trip, with the same bullock effort and the same amount of the farmer's time. Moreover by 1978 it was not necessary to pay the full day rate, since the marketing was generally done at Uncha Pind and Morinda rather than Kurali. Travel time for Kurali had been one day. For Uncha Pind it was a third of a day and for Morinda a half-day. Thus, all things considered, the cost of renting a cart to sell grain in 1978 was lower in absolute terms per unit sold than in 1965.

Finally, the reduction in transport costs was greater still when considered as a percentage of the value of the units being transported. In 1965 the cost of transporting wheat to market was just over 2 percent of the value of the wheat itself. In 1978 it was less than 1 percent. The costs of transporting rice and cotton were even lower. Such reductions are especially significant for the cheaper goods, like fodder and manure, where transport makes up a larger portion of the final delivered price.

Table 5.6

Transport Costs, 1965 and 1978

Cart type	*Price*	*Capacity*	*Useful life*	*Rental rates*
1965				
Gaddā	Rs. 800[a]	8–12 quintals	12–14 yrs.[a]	Rs. .50 per *maund* per day (equal to Rs. 1.34 per quintal)
1978				
High-wheel *rerī*	Rs. 2,000			Rs. 1 per quintal to Uncha Pind (5 trips per day)
		20 quintals	25 yrs.	Rs. 2 per quintal to Morinda (2 trips per day)
Low-wheel *rerī*	Rs. 2,400			Rs. 3 per quintal to Kurali (1 trip per day)

[a]I have not been able to find notes recording these figures and am thus relying on memory.

Next to the oxcart, the most important means of transportation in the village is undoubtedly the bicycle. In 1965 prices varied rather widely in the cities, but in the village only the cheaper bicycles were used, made in Ludhiana as copies of British workmen's bikes. They generally sold for between Rs. 20 and Rs. 30. In 1978 they were the same machines, but they sold for between Rs. 160 and Rs. 300 (the latter being what I paid in Chandigarh, not what I was told in the village), making this the category of transport that had increased most in price. Nevertheless, as noted, many more were owned.

For long distances, buses were most often used. At under Rs. 0.04 per kilometer, the fares had just about doubled since 1965. The major bus service was a subsidized, state-owned corporation. It was supplemented by private local companies that charged similarly low fares. Empty private trucks would often pick up passengers and charge similar rates, especially at night when buses generally did not run. A driver explained that this practice was not really profitable, but it did help pay for the fuel.

The relative and absolute reductions in transportation costs are due only partly to the increased availability of vehicles that followed upon the reduction of licensing restrictions after the separation of Panjab and Haryana in 1967. Another part of the reduction comes from the improvement of roads, and a very large part comes from the state government's policy of establishing many more market and administrative centers. This policy, supporting the greatly increased physical interpenetration of village life and the life of the surrounding region, has led directly to the increased flow of traffic that struck me immediately on my first trip back to the village and on every trip thereafter.

SUMMARY OF ECONOMIC CHANGES

In sum, the basic conceptual elements of economic calculation and the idea of what a market is have not changed. Yet changes in several important market conventions as well as in the relative prices of credit, land, agricultural capital, cattle, labor, housing, and transportation have greatly altered the pattern of economic activity by altering the relative profitability of different courses of action. The changes have at once provided increased economic security, encouraged more development of agriculture, and provided more opportunities outside agriculture.

It would be wrong to say that in 1978 the farmers were more willing to take economic risks. They were always willing to take risks. It was only in 1978, however, that it made sense for small farmers to innovate at the same rate as large ones. They had better credit, equal access to an assured supply of the materials necessary for success, and more reason for confidence that what was being recommended by development agencies would work under local conditions. At the same time, all farmers had protection against declining crop prices in the event of increased harvests. Therefore there was no reason for a small farmer to sit back and wait a few seasons while a larger farmer, who could afford to set aside a small portion of his land to experiment with, tried the new crop or technique. This may seem a small point, but when one considers that most of the land is in the smaller holdings and that new innovations are being offered every year, the multiplier effect can increase by many orders of magnitude the number of years required for an improvement to be completely adopted and its accumulated benefits obtained. In this sense, there has been a quite marvelous fit between the pricing structures of the economy, the biological interrelations within the ecology, and the managerial requirements of the division of labor, and it should immediately be said that this has been no accident. It has been the result in most cases of careful and very sophisticated thinking by government and cooperative officials, often with rural backgrounds themselves, who at last seem to understand and respect the villagers and who provide incentives that will attract the farmers' interest and activity instead of demands that farmers will have to ignore or resist. Between 1965 and 1978, the economy became a much more effective instrument whereby the private economic interests of the villagers were united with the public good of the region—not only in the official view but in the villagers' own view—and this was unquestionably a major source of villagers' satisfaction with the changes.

Above:

Mid-June 1978: The shade of a pipal tree on a hot summer day; the chauk *by the village mill.*

Opposite:

Covered courtyard of Indar Singh's house, from the entryway. Door leads to interior work and storage rooms.

Above:
June 1978, sunrise: Contract laborers of Sweeper caste loading the last of a farmer's compost heap on a rerī *to be transported to fields for new crop of cane. Second compost heap is in background.*

Top left:
June 1978: Using a single electric motor to pump water and to drive a drummy for chopping and mixing fodder (dried lentil plants and wheat straw).

Opposite:
July 1978, rainy season: The new brick and concrete houses along this lane belong to the families of Sweeper caste who in 1965 lived in the only mud brick houses in the village.

Above:
Summer 1978: the village gurudwara. Forward portion is the new rest house.

Opposite:
December 1965: Diwan *in the village gurudwara commemorating the martyrdom of the sons of Guru Gobind Singh.*

6

CHAPTER

The Kinship System

The kinship ideas in use in Shahidpur, and Panjab more generally, did not change notably between 1965 and 1978, and this stability itself is an important part of what allowed the Green Revolution to arise naturally and painlessly from the actions of millions of people responding properly and reasonably according to their own lights in their own familiar contexts and in pursuit of their familiar goals. Yet this stability was not automatic. There were governmental actions that would have radically altered the established kinship conceptions, and if efforts to impose them had been pursued, they could have disrupted or reversed the progress that occurred. Many in Panjab at this moment fear they may yet do so. The conflicts over these ideas are in fact a major part of a continuing and far-reaching constitutional crisis that concerns the relationships between central and state governments, law and custom, fundamental individual rights and national development goals, and the status and nature of personal property rights.

The ideas of kinship, like the key ideas of the economy, are established by convention and maintained by public sanctions. The most general ideas, like the economic ideas of mathematical quantity, are far too abstract and appear in too many different specific applications to be attacked in all their applications at once. But each specific application is supported by specific sanctions and these can be altered.

Just as specific applications of mathematics in the economy are supported by sanctions backing loans or the arrangement of sale and mortgage, so specific applications of kinship ideas are supported by legal records of family rightsholders and, most importantly, by customary law. Customary law is, basically, recorded custom which is enforced by formal courts as law. The present records of custom for Shahidpur were started after the British took over the area in 1848. When recording original land titles, district-level officers were required to visit each village and record the usages pertaining to property rights, especially rights obtained by birth and marriage. Thereafter, these recorded customs were to guide the revenue clerks as titles were changed. When disputes over the changes of the records arose before the revenue officials, they were subject to appeal to the regular judicial hierarchy, which in turn was authoritative for the subsequent interpretations the revenue officers could make. As the courts attempt to interpret and apply these recorded customs, their implications evolve just as happens with law originating in legislation. At the present time, while recorded precedent has a *prima facie* claim to be legitimate custom, the main test is whether the usage has major support among authoritative representatives of the community from which the usage comes—hence in practice the preponderance of weight actually goes to witnesses in each case.

Customary law is especially concerned with inheritance, and hence with the maintenance of the family organization. The main threats to kinship ideas have come through legal attempts, intended nominally to bring about agrarian reform and some social reform, to direct courts to modify the pattern of rights they will enforce in such a way as to depart from customary law in both its established senses: as what has been documented in the past and as what is attested by a majority of people in the community at present. Some of these modifications, as will be described, violate ideas at the core of the entire kinship system. They thus remove the sanction of law from the entire system of derived ideas and place the law, and hence the mechanisms which enforce it, in opposition to the system. When this happens, one is faced with a choice between what is legal and what is meaningful. It is as though, in the economic context, the government solemnly declared that henceforth whenever two odd numbers were added together, the result had to be odd. It is impossible to conform to such a demand; one can only ignore it or resist it. The greater the force with which attempts to impose such demands are backed, the greater will be the resistance and potential disruption.

Kinship ideas differ from economic ideas in that while the ideas of the economy claim on their face to apply to inanimate, objective "stuff"—money and goods—kinship ideas all claim to apply to animate human beings. Further, these human beings include not only those whom the user of the ideas is thinking or talking about, but the user as well. In this sense, the definitions are subjective as well as objective. They define the user (the subject) as well as the object of his action, and by the same token their use implies a commitment on the user's part to accept the relation that he himself assigns in dealing with the person spoken to or thought of. This is one reason ideas like kinship strike us as specifically "moral," as entailing obligations or commitments, while economic ideas do not, even though both types of ideas actually require social enforcement.

The villagers of Shahidpur describe kinship as relations (*risten*) by birth (*sāk*) and marriage (*shādī*). (*Sāk* can also be translated as "relative." In its broad interpretation it would include relatives by marriage; in a narrower interpretation it excludes them. I take it as being a morphophonemic variant of the adjective *sākā*, which has the clear sense of "own" or "true" relative as in the phrase "true brother," meaning brother born of one's own father.) The area that contains this system is the region (*ilākā*) not the village, because every village family necessarily has relatives in other villages. The region in this sense is spoken of as family centered and relativistic, different for every village household.

The most basic ideas of the system, functioning like the basic mathematical conventions and the concept of legal tender in the economy, are the kinship terminology and a general concept of property. Every concept of a specific kinship group consists of a selection of relations defined in the terminology and a particular version of the general idea of property. This is an important point to understand—that "property" in this system is as fundamental a kinship notion as the idea of birth itself and that, in effect, all relations between persons as members of kinship groups are at the same time relations revolving around, and defined by, connections to property. This is precisely why kinship groups are so important in the division of labor and, conversely, it is precisely why legislative attempts to alter property rights have attacked the kinship system as well.

THE CONCEPT OF PROPERTY

The concept of property (*jaidād*) is expressed in the distinction already noted between property that is ancestral and that which is self-acquired. Ancestral property is what sons obtain from their father by birthright. Obtaining it this way imposes the obligation to pass it on to one's own sons in turn, and this obligation has traditionally been sanctioned by customary law. Such property is conceived as something that can be used to earn a livelihood. In this sense a radio or clothing would not be stereotypically offered as examples of ancestral property since they are not productive. Nor would a doctor's skill be ancestral; it cannot be shared or passed on. Land, by contrast, would be presumed to be ancestral. So would a shop, for a shopkeeper. The idea is vague and there are constant arguments over what it actually applies to, both in courts and in less formal settings. But the vagueness is itself important and part of what gives the idea its general integrative power and broad moral force.

THE KINSHIP TERMINOLOGY

The kinship terminology is the set of terms and their definitions that establish the reciprocal relations between a person and his or her relations. It is abstract, but it is precisely this abstractness that lets it serve as the conceptual glue that holds the seemingly more concrete definitions of specific kinship groups in their fixed logical relations to one another, defining the position of the person who views them not as a member of so many completely separate watertight compartments, but as an "I" at the center of a single, coherent system of interrelated groups and groups within groups interacting with others who are equally centers of their similar systems.

In 1978 I did not attempt to repeat the systematic elicitation procedures I had used in 1965 (see Leaf 1972:172–178). Rather I went over the discussion in the book and used the information in conversation, watching for indications that my usages might be odd or confusing. There were no problems; the basic terminology was clearly the same.

There is one main kinship terminology which is used both to describe relations and to address them. This is given in Figure 6.1. There are two minor terminologies, used only for description, which are defined on the basis of the main terminology. These are the expressions that a husband uses for certain of his wife's relations as defined in her use of the main terminology (Figure 6.2), and those that a wife uses for certain relations of her husband (Figure 6.3).

It is important that every feature of the charts have indigenous names and indigenous meanings: the triangles that represent males, the circles that represent females; the lines representing *sākī* (relations of birth), the equal signs representing *shādī*, marriage; the vertical dimensions which were easily described by villagers as up and down, high and low, and represent generations; and the horizontal dimension representing, roughly, the relations near (close to the central Ego) and far. Even the chart itself was spontaneously described as a map or chart (*nākshā*) by those who saw it for the first time after it was set down on paper, who were not in the original group from which I elicited them. The idea of an "Ego" at the center was also readily accepted and used, and there was no problem in drawing it with a circle and triangle, signifying that it could be either male or female.

The logical core of the terminology is the configuration of six terms in the

Figure 6.1
The Kinship Terminology, Main Terms

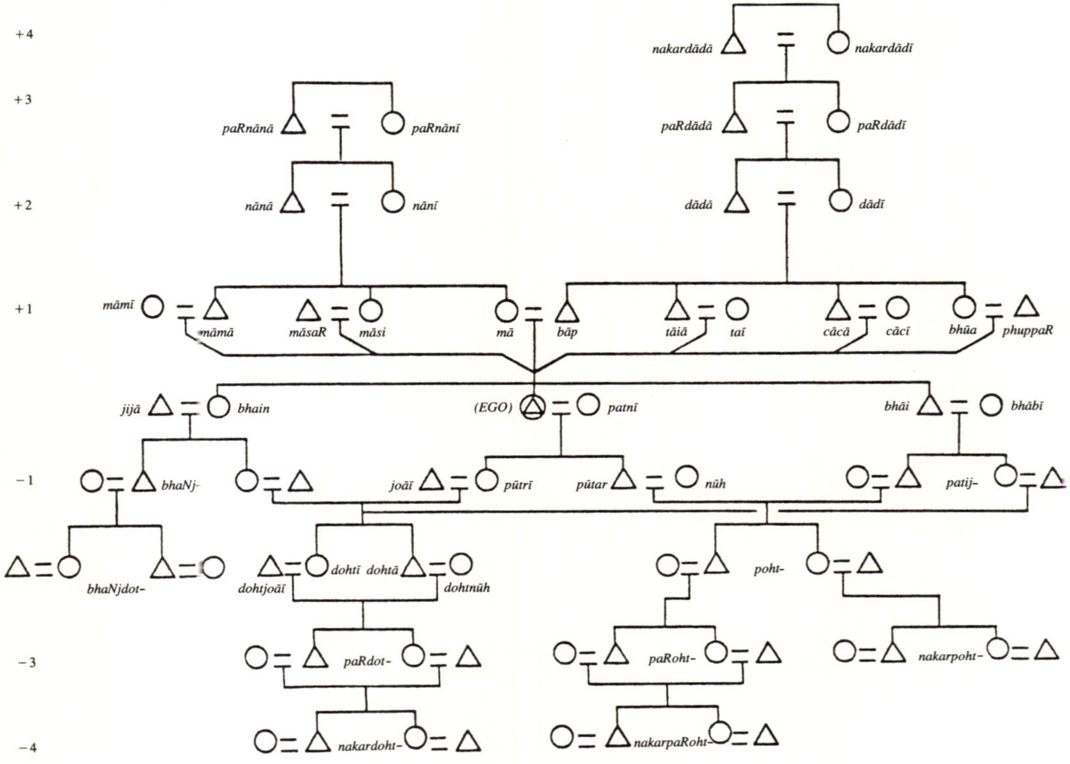

Figure 6.2
Terms for Wife's Kin

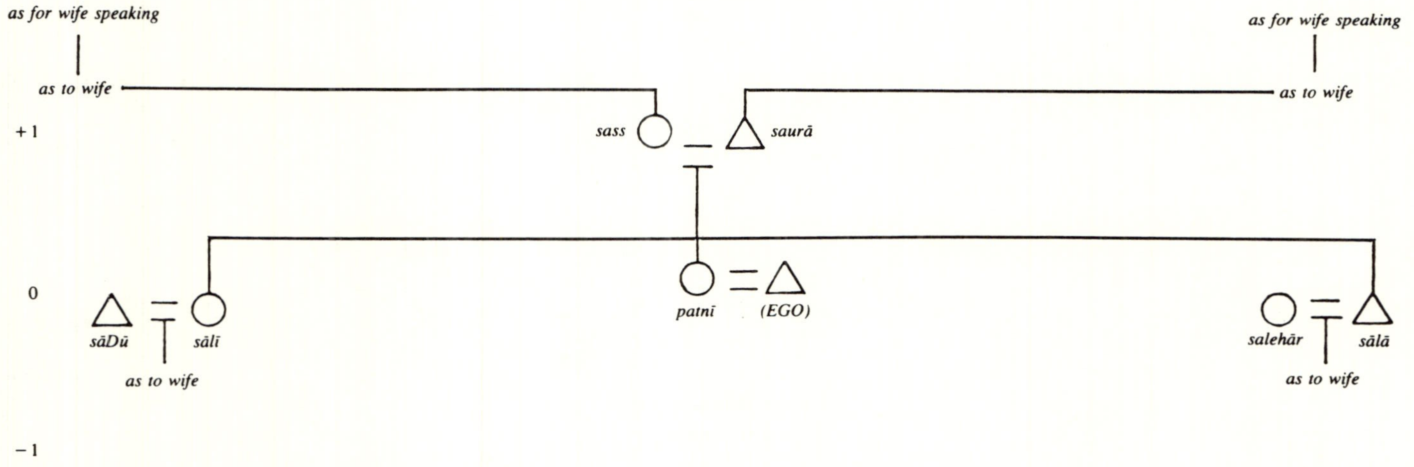

Figure 6.3

Terms for Husband's Kin

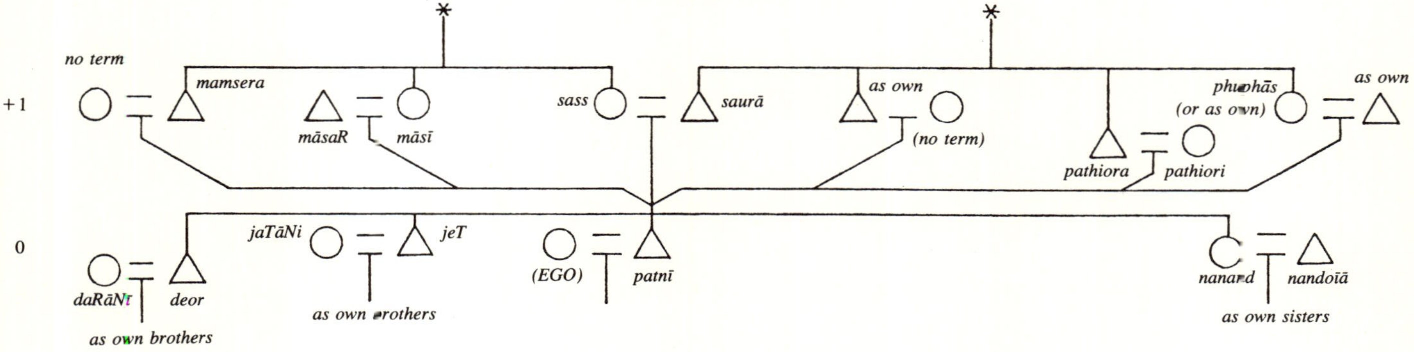

center of the map that are most directly linked to Ego, everyone's "closest" kin: *bāp*, *mā*, *bhāi*, *bhain*, *pūtar*, *pūtrī*. These and only these. It will be tempting for English speakers to translate these as "father," "mother," "brother," "sister," "son," and "daughter," respectively. Yet they are not exactly the same.

In the initial elicitation, these six direct relations served as the framework that was used to obtain the rest of the structure. Since everybody could have these and only these types of direct relations, the elicitation proceeded outward by taking each of the six direct kin in turn and asking what *their* direct kin were to Ego. Thus, taking *bāp* first, the questions were: "What is the father of your father to you?" "What is the mother of father to you?" "What is the brother of father to you?" "What is the sister of father to you?" "What is the son of father to you?" and "What is the daughter of father to you?" In Panjabi, these questions follow a standard grammatical form (which those teaching me the terms prompted me to use) and are quite unambiguous. The answers yielded some new terms and some repetitions of terms already elicited. For example, mother of father was *dādī* and father of father was *dādā*, both of which were new. These were added to the drawing. Son of father (*bāp dā pūtar*) and daughter of father (*bāp dī pūtrī*) were *bhāi* and *bhain* respectively, terms which had already been elicited. In this case (and for all such questions) I asked if they were the *same bhāi* and *bhain*, with no difference, and if the existing representation was good. If it was, as it was in this case, nothing more was added except that I mentally added the new information to the meaning of the graphic form. More often, when a previously elicited term was encountered again, a modification of the chart was required. For example, when the sons of all the brothers of *bāp* and *mā* were seen to be *bhāi*, the converging lines dropping from the +1 generation were added to show this, after those present approved it. In general, my rule was to represent one concept with just one graphic position, but only after discussion to be sure the position was indeed one and not thought of as two distinct positions that merely had the same name.

After going around all of the direct kinship relations in this way, the procedure was repeated for each of the new positions that had been turned up, with continuing modifications of the chart. These were queried in the same way in turn until all the endpoints had been reached and the system was logically closed—and said to be so by those present. This is the only method of kinship elicitation that provides such completeness.

In Panjabi, endpoints are marked in two different ways. Both are quite unlike the way endpoints are marked in English. Lateral endpoints are marked by positions whose definitions include rules for extending the terms to cover an unending series of new individuals of the same type on the same generation. On one's own generation, for example, all males who are not brothers-in-law (*sāDū* or *sālā*) are *bhāi*—including all the sons of all of those on the next generation up, and all the fathers of those on the generation below. On one's father's generation, there are four types of male relations possible, and there are rules to be sure that every male one can possibly encounter can be properly placed in at least one of those categories. There is, thus, no lateral distinction like the English boundary between kin and nonkin. There is no line beyond which people on a generation are not relatives.

By contrast, there are in Panjabi definite boundaries between relations and nonrelations in the vertical dimensions. The endpoints that mark them use definite prefixes. In the male line, the prefix is *nakar*; in the female line it is *paR*. In the initial elicitation when we got to *nakardādā* and I then asked "What is the *bāp* of *nakardādā* to you?" the response was instant laughter and much amused repetition of the question. I still don't know why it was quite that funny, but I was clearly told by everyone that beyond *nakardādā* there were no relatives. Someone may be an ancestor, but they were not relations (*risten*). The same is true for people who are descendants of *nakardohtā* or *nakarpaRohtā*. They are descendants, but not relations. So it is with all the positions represented on the chart as endpoints at either top or bottom.

When all the endpoints were reached and understood, all the implications of the original core ideas were fully spelled out and represented by the graphic forms and the verbal explanations for them. The complete terminological map gives every pair of reciprocal terms used in Panjabi and allows one to predict every possible way each term will be defined by using either other kinship terms or the more-general kinship ideas. (As in English the term "cousin" might be defined alternatively as son or daughter of uncle, son or daughter of father's sister, or as a collateral relation of either sex on one's own generation.)

Figure 6.1 is exactly the same as the village map I drew in the initial 1965 interview except that the names of the relations have been transliterated from the Gurumukhi, the indigenous script in which Panjabi is usually written.

To account for the ease with which the chart was elicited and with which it was recognized by others later for what it was, it should be noted that the graphic blocks of terms in the chart correspond to poetic and grammatical patterns in the terms themselves. All the terms above Ego's generation are in pairs, and each pair is constructed with a unique stem and then either the regular male (*ā* or *aR*) or female (*ī*) ending. As the *nān-* and *dād-* terms designate relations more distant from Ego, they also acquire prefixes that designate progressively greater generational distance: nothing (no prefix), *paR*, and *nakar*. Below Ego's generation, all terms are compounds in sets of four constructed in a regular way using the stems that are recorded and one of four endings. If the relation is a male related by birth, the term takes the regular Panjabi masculine ending *ā*. If a female related by birth, the stem takes the regular feminine ending *ī*. If a female related by marriage, the ending is the term for one's son's wife, *nūh*. For a male relation by marriage, the ending is the term for one's own daughter's husband, *joāī*. Thus, for example, the *pūtar* of one's *bhāi* is *patījā*. His wife is *patijnūh*. His sister is *patījī*, and her husband is *patījjoāī*. Figures 6.2 and 6.3 do not alter this basic logic. They only give additional terms that one can employ descriptively for near kin to one's spouse, as indicated. For address, one would use the *same* terms as one's spouse, or proper names.

The actual terms recorded are the standard names for each position. There are many other linguistic forms considered to be their synonyms, but they do not represent different basic definitions. If one asks what a *gharwālī* is, the answer is *patnī*.

The generative power of the chart is enormous, but a few examples will suffice to show how it works in general and some of the more pervasive, particular implications. Consider the terms *dādā* and *dādī*. The graphic forms show that *dādā* is male, and *dādī* is female, and that both are on the *bāp* (father's) side on the +2 generation—what we would call the grandparental generation. By joining them with both a solid line ("birth") and an equal sign ("marriage"), the map represents the rule that *dādā* can be either the sibling or spouse of *dādī*; and *dādī* can be in turn either sibling or spouse of *dādā*. One person cannot be related to the other both ways, of course, since marriage is forbidden between brother and sister. That is, we would say in English that the *dādā* is either grandfather or great uncle of Ego—but even this would not represent the idea completely, because the English terms still differ in their referential scope.

The logic of this configuration leads directly to the idea of an indefinite extension of the terms outward on the *dādā*'s generation. For example, if the *bāp* of *bāp* is *dādā*, and his wife is *dādī*, then logically all of her brothers ought to be *dādā*. But if they are *dādā*, then in turn all their wives must be *dādīs*. And if all these women are *dādī*, then all their brothers in turn must be *dādā*—and so on forever. When I pointed this out during the initial interview and in subsequent interviews, there was no consternation or any sign of difficulty. Rather, there was agreement—"Just so, you are getting the idea." In effect, anyone who is male on that generation can be *dādā* and anyone who is female can be *dādī*. This is part of the idea of the terms, part of their definition, but it presents a further logical problem. Notice that exactly the same graphic configuration occurs on the mother's side, with the pair *nānā* and *nānī*. The form has the same meaning there, so anyone on that generation can also be *nānā* if male, and *nānī* if female. When I pointed this out, the response was again not consternation but agreement, and this is a point of the greatest generality and importance. There is no sense that a person has to be a relation of one and only one kind. Instead, there are established rules for which of the available choices should be made under various circumstances. If one direct connection is known, the term is chosen for that side, mother's or father's. If two direct connections are known, one on each side, one chooses the term for the side through which the relation is closest. If there are no known connections, the term is chosen by location. In one's mother's village and dealing with kin of one's mother, the mother's-side term is preferred. Otherwise, the father's-side term is to be used.

Sometimes even these rules are not quite enough, and jokes and anecdotes about the possible problems abound in Panjabi folklore. One example of some significance in the recent history of Shahidpur will suffice. Gurdev Singh, a villager in his thirties in 1965, was active in village politics and was a regional supporter of the headmaster of the nearby higher secondary school, the same whose family had originally been from Shahidpur as described in the first chapter. By known kinship links, Gurdev was in fact the *dādā* of the headmaster, but the headmaster was chronologically much senior, as well as more prominent politically and professionally. Accordingly, it would have been something of a joke for the kinship term to be used in public. Thus the term was used only when the two men were alone or among other true relations. In public the younger man always called the elder "Headmaster." Technically, Gurdev Singh was of course also the *dādā* of the headmaster's wife, but here again it was also possible to reckon by generation. Since she was older and embarrassed by the possibility of calling Gurdev *dādā* while he called her by name (he could not say exactly why), she called him by name and he called her by a term like "younger aunt" (*cācī*). The story, told by Gurdev himself, clearly assumes that kinship relations are matters of choice and agreement in a way that is quite different from the American notion expressed in the saying "You can choose your friends but you can't choose your relatives."

Because the upper terms are broad, each of them can have several reciprocals. To find the reciprocal used in any specific case, it is necessary to draw the genealogical links and then read off those links on the chart, taking each endpoint in turn as Ego. If two people are male, and one is the son of the son of the other, then the elder is *dādā* and the younger is *pohtā*. If one person is the son of the other's daughter, then the elder is *dādā* and the younger is *dohtā*.

KINSHIP GROUPINGS

Seen purely as an abstract system, the Panjabi terminology could logically support an almost infinite range of conceptions of seemingly concrete kinship groups. Think, for example, of the groups that can be constructed simply by taking all *dādās*, all *nānās*, all *māmās*, and the like. Out of this exceedingly large range of logically possible groups, the explanation for the way widespread consensus forms on just a few of them lies in the establishment of specific institutional mechanisms that enforce some ideas but ignore or set aside others. Every idea of a major social group current in the village and area is derived from the basic ideas on the one hand and supported by specific sanctioning institutions on the other.

The groups most often spoken of can be described in a descending order of their relative scope. The largest would be the totality of all people. This may sound too abstract to excite much personal interest, but it is exactly what gives such pervasive scope to threats to disrupt the basic kinship concepts. Next in order are the concepts of *jātī*, *got*, *pattī*, family, and finally the smallest unit would be the individual as part of a kinship system. The "largest" groups are not the most important on a day-to-day basis.

The totality of all people is a kinship group in the sense that the main kinship terms are proper forms for universal polite address. A male of one's own village will be called *bhāi* if he is of one's own generation. If he is of one's father's generation, and older than one's actual father, he should be called *tāiā*, and if younger *cācā*. If he is of the next generation, he should be *dādā*, and so on. For one's mother's village, one uses the mother's-side terms. For other places, one treats people as for one's own village. The terms for those below Ego's generation are not normally used for address. In the absence of close known relations, proper names are used to reciprocate the superior terms. In this context, the idea of property that is assumed is the world as a landscape of villages, of different places to which different people are tied. The world in this sense is the patrimony of all those living, to be passed on to those to come.

The *jātī* is probably the Indian kinship group best publicized in the West. For address within the *jātī*, the same rules apply as for people in general. In addition, *jātī* names may themselves be used vocatively. *Jātīs* are distinguished from each other in just two ways. First, since all marriages should take place within one's own *jātī*, the terms for affines logically can only apply to those whose *jātī* is the same (remember, however, that these affinal terms are not used in address). Second, those of the same *jātī* are considered to have the same *kind* of property and hence the same occupation, in a distinct region (*ilākā*). *Jātī* origin stories follow a common and simple pattern that illustrates what *jātīs* are thought to be. The present group is always represented as descended from a past group who came into possession of that particular type of property in that particular region through some historical event, and the property is described as having been passed on ever since through inheritance. For landholders, the story usually is that they came into the area as a military body, defeated the former inhabitants, and took over their land. For other groups, stories vary. In some cases, a *jātī* is held to be the group that was defeated and therefore left landless. In other cases, a *jātī* is described as an ethnic group that came into the region at some point in time and was given a certain task (by a ruler, by the dominant group, or by persons unspecified). Accordingly, if two people in a region have the same occupation, this is *prima facie* (but indefinite) evidence that they are of the same *jātī*. *Jātīs* are virtually never represented as descended from a single common ancestor, and are never represented as a kinship group within an occupational group. The stories always assume just one group of descendants with one occupation per region.

I should repeat what has already been said in discussing the division of labor: everyone knows that many actual occupations (like doctor or soldier) cannot be thought of as based on property and many types of property (like shares in a corporation) cannot be thought of as located in a region. Such occupations and such properties are simply not mentioned in the stereotyped stories and are not normally brought up in conversations about *jātī* activities. They are part of different universes of discourse.

As kinship terms are considered polite forms of address in general public settings and suggest a broad commonality of interests in property, *jātī* names are considered appropriate for address in business contexts and suggest a complementarity of interests, although noncaste occupational names like doctor, headmaster, and driver are also commonly used in the same sense.

Sanctions supporting consensus on the concept of the *jātī* are rather complex, but do not in any way include *jātī* bodies that oversee the activities of their members and safeguard ritual exclusiveness either at a regional level or within villages, as has sometimes been suggested. There are no such bodies in Shahidpur or anywhere around. Rather, the main sanctions at the present time are actually legislative measures intended to remedy *jātī* discrimination. These include most notably provisions for reserved seats on panchayats and on legislative bodies, reduced tuitions for "backward classes" in schools, and the 25 percent reservation of job application positions in government employment. An additional important sanction lies in the practice of recording *jātī* names as part of personal identifications in the genealogical record that supplements the official record of rights. The idea of a *jātī* might also be of increasing importance in national politics, as the central Congress party has more and more been seen as an alliance of industrialists and "Harijans" against "Jats" (and similar farming groups). Finally, a subtle but persistent sanction lies in the convenience of *jātī* as the best-recognized set of socioeconomic labels with accepted meanings (however vague) to describe village occupational groups, both for managerial planning and addressing people to initiate discussion of various contractual relations.

The clan, *got*, is a descent group in a strict sense. Like *jātīs*, *gots* are named. In contrast to the origin stories for *jātīs*, those for *gots* generally represent the *got* as descended from a single founding ancestor or family, although the name of that ancestor or family is often unknown or indefinite, as are the links to him. *Got* names are arbitrary and usually contain no occupational reference, even though it is conventionally stated that each *got* is, necessarily, within one and only one *jātī*. Unless there is reason to believe otherwise, *gots* with the same name in widely different regions, or in different *jātīs*, are presumed to be not really the same. Marriage is prohibited within the *got*. In fact, it is prohibited within the "entire four" (*pūrā cār*), which are the *gots* of one's father's father, father's mother, mother's father, and mother's mother. Because marriage implies that men stay in place while women move to their new husband's area, *gots* are not thought of as related to regions in exactly the same way *jātīs* are. Only the men are localized; the women would be assumed to be dispersed. If one asks, for example, how many *gots* there are in Shahidpur, the *gots* referred to must be of men only; their wives necessarily belong to different ones, and the *gots* of the women who have come to a village through marriage must logically be much more numerous than those of the villagers born locally.

As the idea of a *jātī* is connected with the idea of a common and unique kind of property, so the idea of a *got* is connected to the idea of a subset of its *jātī* property which is to be found in a specific group of villages. In the middle of the last century, when the British administrators established the revenue records for this area, members of different *gots* were allowed to attest to different patterns of customary law. People of several of the *gots* of Shahidpur recognize common kinship with like-named groups in nearby villages. The main group of Jat farmers, for example, identify their *got* as being the same as the major *got* in three nearby villages which are said (in the other villages as well) to have come from Shahidpur.

Gots are recorded along with *jātīs* in the village genealogical records. *Got* names are part of each individual's personal name. Except for its use in marriage regulation and personal identification, the idea of a *got* is not considered particularly important.

A *pattī*, meaning a share of property, is a group of "rightsholders" descended from a common ancestor who have a share of the total properties of their *jātī* and *got* within a single village. *Pattīs* are usually referred to by a nickname based on some outstanding characteristics of its prominent members—like the "fools' *pattī*"—or simply the *pattī* of so-and-so. A *pattī* cannot extend beyond a village, although a village *pattī* may have members not in residence. In the case of the *pattī*, the ancestor is usually a fairly recent figure and often known by name. If a *jātī* is represented in a village by only one family or a group of families with a known common ancestor, then that group will be a *pattī* and whoever founded it will be the founding ancestor. If a *jātī* is represented by several *gots*, then each *got* might be a *pattī*. Or if a *got* is large, it might be divided into several *pattīs*. In Shahidpur, each Brahmin family, being of a distinct *got*, is also a distinct *pattī*. The one Water-carrier family is a *pattī*. Among Harijans, each of the separate *gots* was recorded as a separate *pattī*. Among Jats in 1965 the *got* represented by a single family was not a *pattī* since it had no land in the village. The large *got* was divided into six *pattīs*. In my later study, the four small *gots* were represented again by one family each, but two of them did own property, and were recorded officially as distinct *pattīs*. Most families still belonged to the one old and large *got*, and the same six *pattīs*.

Pattīs are the basic organizational unit of the village genealogical record: families of each *got* in each *jātī* are grouped together into legal *pattīs* which correspond closely, but not exactly, to the groupings villagers will list on their own. In the old records of customary law in the village, *pattīs* were often mentioned and laws could vary by *pattī*. In 1965 farmers said that, in the past, rights of preemption had extended to the *pattī* but these were no longer supported by courts. By 1965 members of *pattīs* were just felt to be "closer" than others in the same *got* or *jātī*. Between 1965 and 1978 there had been no further change.

The idea of a *parivār*, family, is by far the most important of all the applications of the basic kinship ideas. It is sanctioned officially in legislative and customary law and in local practice by all the pressures and constraints that lead to the self-perpetuation of the organization of the system of management around physical households and those who reside together in them.

As a *pattī* is not quite a segment of a *got*, so a family is not quite a segment of a *pattī*. As noted, in 1965 a *parivār* was described as a group of kin cooperating around some common property. Close kin without shares in the same property were not in the same *parivār*; distant kin who were involved were in the same one. Servants in a house, however, were not part of the *parivār*. Women married in, who are not (stereotypically) rightsholders and therefore not in the *pattīs*, are nevertheless part of *parivārs*. Indeed, I think villagers would agree if I say they are its heart. This is related to a second point.

The property of a *parivār* in this definition is not assumed to be just the standard property of the *jātī* it belongs to. If it is a Jat family, the property is not only land. If it is a Mason family, the property is not only the Mason's tools. When speaking of *parivārs*, "property" means all the substantial goods in the house and the house itself; the term "house" (*ghar*) itself is in fact commonly used as a synonym for *parivār*. Cooperation in such property consists in all the daily tasks that make for physical survival—food preparation and processing, making and maintaining clothing, cleaning, budgeting, seeing to the education of children, tending to the ill, having babies, hiring help, feeding help, and so on. This is the concept of property courts support, not the concept associated with the concepts of *jātī*, *pattī*, or *got*.

It is considered that no one can live comfortably in a village without a *parivār* because village living takes too many skills for one person to master or have time to execute alone. This is why being an "orphan," a person without a *parivār*, is considered such a great misfortune. It is to have nothing, as well as no one. It should be noted that there is no concept of minority in this context. A *parivār* is not primarily the main married couple, or even the adults of the house. The smallest child is an equal part, and will contribute in proportion to his or her ability.

In addition to the requirement that one marry outside one's parents' parents' *gots*, the relations among *parivārs* are shaped by the rule that one must marry outside one's own village. This means that every boy's own village is different from his mother's village, which in turn will be different from his father's mother's village. Every girl's village is different from her mother's village and from her father's mother's, and will eventually be different from her childrens'. This is what links every village household to families in many other villages, in a pattern that constantly shifts through the generations.

Within each village it is assumed that *parivārs* (at least those of the same *jātī*, if we are talking in stereotypes, or of the same economic category if we are being more sophisticated) are in competition with one another—often bitter competition. But since the *parivārs* connected to one's own by marriage are in other villages, they are assumed to have no real basis for such competition and can be steadily supportive. Thus each village family expects its related families in other villages to support its local efforts—sometimes economically, often in such matters as finding employment for children and giving them places to stay while attending school, and almost always in ceremonial and social support (attending weddings, supporting court appearances).

It is often said that one's mother's family is just as important as one's father's family and in later life one's wife's family is just as important as one's own. The term for the family of one's spouse is *sussrāl ghar*, from the kinship terms for parents of spouse (*saura* and *sass*). The term for the family of one's mother is *nānke* (from the term *nānā* plus the possessive *ke*). The natal house of a man is his *dādke* (meaning "house of his *dādā*," in the same way as *nānke*). For a woman, the term for her natal house is *peoke* (from *piu*, another term for father, and the possessive). The latter distinction indicates the difference between the importance of inherited rights for the man as against personal obligations for the woman.

As the smallest unit in the kinship system, the person is not defined by any one role or relation, or by any one exact concept of property. Rather, the person is defined by a sequence of relations occupied through life, and by a sequence of relations to property within his or her world, *jātī*, *got*, *pattī*, and *parivār*. This sequence is fundamentally different for men and for women, but it is basically the same for all men and for all women, regardless of other identifications (such as *jātī*). It defines the basic female and male life cycles.

LIFE CYCLES, FEMALE AND MALE

The same sanctions, legal and practical, that support consensus on the several group definitions also determine a standard expected life career for a man and for a woman. In addition, the life cycle ideas are promulgated and supported by an extensive series of rituals. These include life cycle rituals proper as well as rites of passage: ceremonies marking naming, marriage, birth of children, and death (see Leaf 1974 for a description of the wedding ritual and its symbolism). There is also an annual cycle of rituals that call for members at various points in the life cycle and that therefore describe implicitly the attributes of those points (symbolizing the passage through life by the passage through the agricultural year from the bare ground of the hot season to the harvest of the winter crop) (Leaf 1972:191–193). Most of these rituals focus explicitly on women, and the woman's life cycle is in general much more clearly represented, but the man's cycle is inevitably implied. There is also an extensive folk literature, consisting mainly of poetic folksongs, widely known, often described in local literary works in Panjabi, and often performed by professional and semiprofessional groups.

A woman begins life as a propertyless but loved child in her own village, and ends as a powerful manager of an extended family, her sons and their wives and children, in the village of her husband. The first of the women's rites is *tīj*, at the beginning of the summer rains. It calls for unmarried girls or girls just recently married, and portrays unmarried girls wholly within their natal villages as spending their time among their village girlfriends, doted upon by their parents, and doing favors for their brothers. The next stage is given in the wedding rituals, which portray the bride as being sent on her way into adult life by her parents, advised by an elder sister, a free gift of her father to the family of the groom whom they respect, forced to leave the warmth of her own home and face the domination and exploitation of her new mother-in-law.

The second annual ritual occurs in the first harvest season and portrays the young wife as dividing her time between her natal village and her husband's village, cared for and supported in each by the men in the other, being given material goods by her husband, and giving gifts to the village priests to show respect and ask for blessings. The main theme of the ritual activity is expressing hope for children. In the third annual ritual, taking place as the winter crops mature, the woman with young children is portrayed as completely attached to the village of these children (hence of her husband), being concerned for the welfare of all the village children, and also being a manager of household servants. The next stage is again in the wedding rituals. It portrays the mother of the bride as sending her daughter off for her own good to begin life in another village, but also to clear the way for her own sons to obtain brides. (The convention represented in the wedding rituals is that the daughters are married first from elder to younger, then the boys in the same order, so long as there is property enough.) And finally the wedding rituals portray the mother of the groom as a distant and more or less terrible mother-in-law, running the household and bending the new bride to its discipline in the interest of her son.

For a man, there is nothing like the radical transformation from being the innocent on a swing in one village to being the terrible domineering parent-in-law in another. A man begins his life as he ends it: as one property owner among many in one village family, who has to do his share or be forced out.

The transition from bride to the senior woman whose power and importance are symbolized in the figure of the feared mother-in-law is the result of two natural processes, set in motion by marriage and shaped by the basic concepts of the family. First, the physical strength of her *sass* inevitably wanes, and the younger woman takes over more and more of her duties. Second, while this is going on, the younger woman's own children, especially sons, are being born and growing up. Within the logic of the property rules this creates an automatic increase in her personal importance as well.

It is commonly said in Panjab that while a man has a legal right to his ancestral property, the relationship among men with such rights is strictly "you scratch my back, I'll scratch yours." Everybody is expected to contribute his share to maintaining the whole, and receives no consideration if they are unable to do so. No one can afford a co-sharer who loses his share, for his share in the rest is still proportional. Cooperation, thus, can easily turn to conflict. By contrast, it is also said in Panjab that while a brother must protect his sister, and a husband should provide for a wife (provided she is not childless, for a reason that will be clearer in a moment), a man's most sacred moral obligation is to his mother. The reason is that she has devoted herself (stereotypically) unselfishly and unstintingly to the protection and care of her son and has, in later life, not only nothing else but no one else. The well-known Indian definition of the cow as a sacred animal actually fits into this context, not that of the ecology. The reason usually given for not killing a cow, or even injuring it, is that it is "like mother." Such an explanation contains a kind of hidden argument, so convoluted it can almost never be brought out in the open. It is as if one asks rhetorically "and if we do this just because a cow is *like* mother, what do we do for mother herself?" To challenge the idea of the sacred cow would not be to challenge ecological practices, but rather to challenge the sacredness of this kinship obligation. This is a challenge for which it would be very difficult to find support.

As a man becomes a husband and then has sons in turn, he gains a necessary measure of security for his property, but in equal measure he loses autonomy. Sons, as co-owners, will maintain what is his by maintaining what is and will be theirs. This is why a wife must have children, especially sons, and cannot be kept if she does not. Without heirs, neither husband nor wife would have a secure future. Nevertheless, heirs bring a man a diminishing share to deal with in his own right. His wife, however, is in exactly the opposite position. The more sons she has, the larger is the total share of the family estate that she has the first moral claim to. With three sons, a father is a one-fourth partner in his property but his wife is the mother of owners of three-fourths of it. An old father is said always to be respected, but often because of this respect you will find him sleeping at night in the fields by his well, leaving the house to his son or sons and their families. But as long as she lives, his wife will be within the house and central to its operations.

With respect to legal sanctions, it is a point of some importance that the *parivār* itself has never been recognized in law or custom as a rightsholding body apart from the individuals that make it up. The men have their rights of ownership, supported by both custom and legislation, and women have rights in maintenance. A girl has such customary rights in her father's personal property, while a wife has them in her husband's. If a woman is widowed with sons, she has a right to maintenance out of their holdings while she lives and may be recorded as the owner holding the land in trust for them. If a widow has no sons, she will be recorded as the owner of her husband's estate, although upon her death the land will go to her husband's near kin. The village gurudwara in Shahidpur was built at the bequest of the widow of a farmer. She had held his land while she lived but he had no relatives in the village for her to pass it on to when she died.

The life cycles as described in the village up to 1978 represented a balanced and coherent system of roles that a person could use as a basis for consistent and reliable calculations of what was in his or her own interest and in the interests of those upon whom he or she depended most.

At the present time, the life cycle ideas and the reciprocal relations they generate loom far larger in individual career planning in the village, and in South Asia generally, than do their counterpart ideas in, for example, the United States. The reason lies not only in their coherence and elegance, and not only in the positive support governments have given through the enforcement of customary law. They lie at least as strongly in a long and deeply ingrained history of things the governments do not do; areas they stay out of. There is no alternative system of even remotely similar scope for providing personal security in old age or misfortune. There is no general system of governmental subsistence maintenance. There is no unemployment insurance, no welfare system, nothing like aid to dependent children or free food programs, nor are there even widely available investment opportunities more attractive than individual household productive units. The life cycle ideas permit one to design a secure and useful life by relying on kinship relations; the history of institutions that make up the context of kinship provides villagers no widely recognized alternative that would let them attain the same important goals by any other equally reliable means. Governmental attempts to alter kinship ideas have not taken this into account. Their authors have in effect attempted to undercut the usefulness of the traditional ideas for those who now rely on them by altering what had been supported as customary law, but they have not attempted to step in to provide alternative bases either for personal security or for allowing individuals to define themselves as important to others.

THE MAINTENANCE OF STABILITY

The vulnerability of kinship concepts to changes in laws regarding property comes directly from the fact that the legal records that provide the basis of title to land and related major properties are virtually the only mechanisms that produce public evidence of kinship relatedness. For example, despite the nominal importance of the concept of "birth," no separate birth records of any kind are kept. There are no documents, legal or customary, similar to birth certificates or baptismal certificates that authoritatively connect a given individual to any specific parents. Similarly, marriage certificates are not required to make a valid marriage. Marriages may be legally registered, but this is seldom done. Houses are not titled, as noted, and *eo ipso* family members are not recorded as legal co-owners or partners in them.

Within the framework of the Indian Constitution, three laws have been especially pertinent to village kinship organization. These are the Panjab Security of Land Tenures Act of 1953, the Hindu Succession Act of 1956, and the Panjab Land Reforms Act of 1972. The Hindu Succession Act is an act of the central parliament. The other two are acts of the Panjab Legislative Assembly.

The constitutional issues that frame these acts revolve around two major points: the relations between central and state governments and the right to property.

The relation between the central government and the states is simplest. Basically, the Indian Constitution formally assigns to the central government all matters pertaining to manufacture, interstate commerce, banking, and international trade. The central government regulates and collects taxes from each of these activities, as well as income tax and certain sales taxes. It cannot tax agricultural land or the inheritance of agricultural land. The Constitution assigns to the states all matters pertaining to agriculture, land, and water. This includes the right to collect land tax. Joint responsibility is assigned for social and economic planning, and electricity. Powers not assigned to the states are reserved to the central government.

This separation of powers works together with a broad customary and legal distinction between towns and villages. The difference is not one of size but of type of activity and form of government. Towns are governed by municipal committees, villages have panchayats. In towns, there are markets, factories, branches of banks, and in general the sorts of concerns for which the central government is responsible. Correspondingly, there are usually courts, police, and central agencies for collecting income, excise, inventory, and market taxes.

In villages only land tax is collected and the only officials normally found are the revenue clerks who supervise its collection on behalf of the state government. In effect, this gives the states a primary interest in the welfare and prosperity of villagers. The central government is oriented mainly toward towns. Within the national framework, the central government and the Congress party that dominates it predominantly reflect the interests of towns; the states and the locally based parties that are prominent within them reflect village interests.

Evidently, this basic division was intended as a system of checks and balances. It has been considerably clouded, however, by the history of basic individual rights.

The Constitution originally guaranteed about eight "fundamental rights."

Of particular centrality to land reform is Article 31. As originally written, it provided that "no person shall be deprived of his property save by authority of law." The First Amendment Act of 1951 added Article 31A, which states that no law providing for, among other things, "the acquisition by the state of any estate or of any rights therein or the extinguishment or modification of any such rights . . . shall be deemed to be void" on the ground it abrogated either the rights under Article 31 or the rights to equality before the law (guaranteed in Article 14) or to protection of freedom of speech, residence, and property guaranteed under Article 19. "Estates" in this sense refers to a "revenue estate," a taxed unit of property, and would include the ancestral land of a *parivār*. The Fifth Amendment, in 1955, added that such deprivation must be for a public purpose and that compensation must be provided, but also that no law should be deemed unconstitutional on the ground that the compensation was inadequate. It further provided that laws that did not transfer property to the states or to corporations owned by states were not deemed to "provide for the compulsory acquisition of property, notwithstanding that it deprives any person of his property." State laws that claim the protection of this provision have to be approved by the President of India and are entered in a constitutional appendix, the Ninth Schedule. Both the Panjab acts are in it. This was the constitutional position in 1965.

By 1978, however, the Twenty-Fifth Amendment Act of 1971 had added, in Article 31C, that no law aimed at securing a set of "directive principles of State policy" could be declared unconstitutional on the ground that it abridged rights under Article 13 (that no laws should be inconsistent with fundamental rights) or Articles 14, 19, or 31. The directive principles had been there before, but had been subordinate to fundamental rights. They included the organization of agriculture and animal husbandry. (See Raman 1977 for a discussion of the concurrent history of "basic rights" in this context.)

The land reform acts reflect a long history of ideological development within the Congress party before independence. The Congress party had identified British colonialism both with large landowners and with capitalism in a system wherein the British used rent-farming to maintain their power and control the rural poor but thereby bought political stability at the cost of general welfare and, ultimately, of economic development. It was held (with much justice) that such rent-farmers had no real incentive to develop their land and did not even understand what it needed. Instead, they extracted as much as they could for maintaining lavish establishments and attempting to buy ever more influence and favor. At the same time, the peasants had no incentive to invest because they could barely survive on what they were left after rents, and if they did invest and raise productivity, profits would only go to the landlord anyway. The Congress party identified its own position as anti-colonial and as anti-capitalist. The term "socialist" was construed as having both implications.

The Congress party's intent was to reverse the British policy by confiscating the land of the rent-farmers above an amount they could actually farm themselves and reassigning it to the occupying peasants, the rent-paying tenants. The aim of this policy was to provide assurance that the power base of British colonialism would be destroyed, thereby securing the new system, and to assure that the owners of the land would be motivated to develop it and thereby increase both their own economic productivity and the general welfare. (For an extended history of this development from the Congress viewpoint, see Malaviya 1954.)

Consistent with the national Congress program, the Panjab Security of Land Tenures Act of 1953 had been intended to eliminate large absentee landowners. Specifically, it was meant to give "security" to hereditary "occupancy tenants" of such landowners, the "true tillers of the soil" in prevailing theory. Security took the form of the right to purchase the land outright or rent it at controlled prices. The law allowed each landowner to select and reserve for personal use thirty "standard acres." A standard acre here was the same as in the consolidation procedure: an actual acre of land of average yields, proportionally less area if the yields were greater than average, and proportionally more area if the yields were below average. Anything over the reserved area was declared "surplus." Tenants had the first option to purchase. If they did not want to buy, they could continue to pay rent at a legally set rate. If there was no tenant, the state revenue authorities would arrange for the land to be put up for sale to anyone else, at the same prices, provided only that such land, when added to what the buyer already owned, did not exceed thirty standard acres. The proceeds would be turned over to the owner upon completion of the transaction. He was required to accept payment in installments if necessary. When the act was first passed, the sale price was to be the average price paid for similar land in the area over the preceding ten years. Because of the steady rise in land prices, this was bound to be below the current market price at the time of the sale. In 1955 the act was amended to make the price two-thirds of the average.

While this may sound straightforward and clear, two major problems lurked just beneath the surface. The first concerned what a "landowner" was, the second concerned incentives to improve production. Could a "landowner" be any person, male as well as female, adult as well as child? Or was it just any male? Or was it just an owner of record? Or was it, as the text suggested at one point, the family? The consequences of these various interpretations for the total area an actual working family might finally retain could be extreme. A family of ten or twelve might commonly have only one owner of record. If all family members could be owners, the upper limit of their property could be 300 acres. If only men could be owners, following traditional concepts of rights from birth, the amount might be 120 or 150 acres. But if only the owner of record (or the family) was considered, the limit would be 30. A series of interpretations of "owner" had the effect of gradually lowering the de facto limit while extending the proportion of the total population the act affected.

Apparently there were no major challenges as the law was initially applied to the largest landholders, in the first few years after its passage. The first serious test of its constitutionality was *Jagan Nath and Others* v. *The State of Punjab* of 1960. Jagan Nath alone had been the owner of record. The revenue officers who applied the law in this case had allowed him only thirty standard acres. Jagan Nath objected on the grounds that his was a Hindu joint family, so all of his sons were owners whether their names were recorded or not. The lawyers for the state argued that this objection had no force because the Hindu joint family was a "person" within the meaning of the law. The court rejected the argument, saying that the idea of corporate family ownership had never been a principle of Indian law. "There is, in the circumstances, no force in the suggestion that the joint Hindu family is a juristic person or should be so treated for the special purpose of the Panjab Security of Land Tenures Act." Therefore this could not have been the intent of the legislature.

The decision in favor of Jagan Nath permitted a relatively high limit on family holdings but was not the end of the matter. It raised the possibility of a discrepancy between the rights of members of a "Hindu" joint family and the rights of members of a Jat family under Jat customary law. Since the latter denied sons the right to demand a division of the property against the father's wishes, there was a clear possibility that Jat families could be held to the limit for one person while Hindu families could be held only to the limit for as many males as were in them. To avoid this (in part at least), the legislature in 1962 inserted section 19E, overturning the Jagan Nath ruling. It provided that where, immediately before the commencement of the act, "a landowner and his descendants constitute a Hindu undivided family, the land owned by such a family shall, for the purpose of this Act, be deemed to be the land of the landowner and no descendant shall . . . be entitled to claim . . . he is an owner in his own right." This had the effect of treating all families on the Jat model.

The new section was immediately challenged on several grounds, including sex discrimination and the fact that it made the head of the family the absolute owner, thus depriving other members of their rights discriminatorily. Both these arguments were rejected on the basis of Article 31A of the Constitution and because the section was simply defining what was meant by "owner" and was *not* otherwise affecting the rights of other members of the family in what was owned under custom or Hindu law. Whatever relative shares were held before land was declared surplus would exist in the same proportion in the reserved amount after the surplus was sold and in the income from the sale or rental of the surplus portion. Thus the law did not affect family *structure*, although de facto it substantially lowered the ceiling of possible landownership. This was the situation in Panjab when my first study began.

The lower ceiling made the issue of incentives to improve production especially critical. If a person had only thirty standard acres and a large family, it would be very reasonable, for example, to try to improve output by digging a well. This would, however, raise the value of the land in terms of standard acres. Could a portion then be confiscated in order to bring the owner back down to the ceiling? If so, it would hardly encourage investment. Just such an instance came to the Panjab High Court in 1966 and the court held that this was not the intent of the law. The ceilings were not permanent; they were pertinent only when the law went into effect.

Villagers in 1965 were acutely aware of the act and its interpretations. No one held thirty standard acres, so all were "smallholders" and therefore exempt from its provisions. Nevertheless, because of this act owners began to be very careful to rotate tenants so that no one could be recorded on the same field for several years in a row (this could create a presumption that they were hereditary "occupancy tenants" within the meaning of the law, rather than nonhereditary "tenants at will"). At the same time, village families were mortgaging their local lands to purchase land forced on the market by the act. Thus the act was contributing to the relatively large Jat out-migration.

On the whole, between 1964 and 1966, the threats to basic kinship ideas posed by the Security of Land Tenures Act were considered indirect (at least after Jagan Nath), and villagers were taking wary advantage of it. The Hindu Succession Act, by contrast, posed a much more direct threat. But in this case the threat was relatively easily diverted.

The Introduction to the act gives its assumptions:

> The Hindu Succession Act has bestowed upon hitherto helpless Hindu females who had no right to "inherit" that property of the husband and take it absolutely. They had only a lifetime estate and they were always dependent on and at the mercy of Hindu males . . . the Hindu Succession Act will help the Hindu females maintain their position in the Hindu Society and will no longer be dependent on the males.

It was felt in some central government circles that the lack of legal rights in property for women had led to women being exploited by men. If they could be given rights on an equal footing with men, the women could assert their independence and the abuses would be eliminated. The problem was how to bring this about legally. A most important component of ancestral property is land, which is beyond the purview of the central government. Yet there is no real reason to doubt that the proponents of the legislation wanted to affect landownership. The solution was to declare that the act was "to amend and codify the law relating to intestate succession among Hindus" (Preamble). Since traditional kinship rules left no scope for choice, there was never any need for "testaments." So virtually all succession was "intestate" in a strictly literal sense. At the same time intestate succession was a new topic of law and therefore fell under the authority of the central government.

The act could not reduce the shares of sons. Since these clearly were based on birthright and not the death of the father, the idea of a testament could not be stretched to cover them. But it did give women a share equal to their brothers' in their father's portion of the estate of their natal family and equal to their sons' share in their husband's portion of the estate of the family they marry into. Hindus were defined to include Sikhs, which must have put a special point on the issue of joint ownership in Jagan Nath's case (since a very large number of Jats are also Sikh), although it was not directly mentioned in the decision.

In 1964 the villagers did not share the view that the act would cure a major social evil. The first response was that married women who received such rights unknowingly promptly began to register petitions waiving them. The second response, becoming well established throughout the area, was that men began to write wills and register them, even though the wills uniformly followed the traditional rules. Thus the property was no longer intestate.

From the village point of view, the law had confounded the basic complementarity of roles that both the male and female life cycles were built upon. For men, it created an automatic conflict of interest between the girl's natal family and her husband's family. It was certain that a woman could not physically farm the land herself (and still run a household and raise children), so someone would have to farm it for her. Who would it be? Her brothers had to fear that her share in their traditional holding would be turned over to her husband or someone from his village; her husband had to worry that her share in his holding would go to someone from her brothers' village. In either case, local control and the ability to coordinate the use of that land with the rest of the family's activities would be lost in a way that was quite different from what happened when ownership stayed within the family or even the village. Moreover the law eliminated the possibility of eventual recombination of inherited portions within the main holdings and thus assured fragmentation.

From the woman's point of view, the first problem with the law was precisely that it made her the focus of this potential conflict, suspect by both the parties she most relied on. Second, the amount of land she would receive under the act (especially after land reform) would not be worth giving up the possible gains from her moral position as a dependent in traditional terms. In the accepted framework she could not have both. The law put her in competition with men but did not give her the means to win more than she would lose. Accordingly, the final outcome was the only one that was reasonable, not only for what each person involved would see as his or her private good but for what they would necessarily see as the general good of society as well (conceived as a system of families).

By 1978 the Land Reforms Act of 1972 had raised a threat to basic kinship concepts that was at once more pointed than that of the Security of Land Tenures Act and more closely tied to viable sanctions than that of the Hindu Succession Act.

The Land Reforms Act had been adopted during the official "Emergency," when the government of Prime Minister Indira Gandhi assumed extraordinary powers. It was drafted in the central government and was adopted in Panjab (according to officials I spoke with in 1978) only after intense central governmental pressure. It was not technically evaluated before adoption and local support came mainly from representatives of "landless groups." It was recognized both locally and nationally as "anti-farmer." Its proponents often defended it (and similar acts in other states) by identifying peasant farmers as "kulaks" in order to suggest that they were both rich (hence not in need of consideration) and inherently opposed to progress. In consequence, it was justifiable to deal with them harshly, on the Soviet model.

The act abandoned the concept of standard acres. Four major classes of land were defined and absolute ownership limits were set in each class. For "land under assured irrigation and capable of yielding at least two crops a year," as in Shahidpur, the limit was seven hectares (17.9 acres). Each landowner could reserve this amount for himself and a like amount for each *adult* son, provided the total for each did not exceed the allowable limit. Minor sons were excluded entirely, although the landowner could reserve an additional fifth of the ceiling amount for each person in his immediate family over five, up to three-fifths. Rather than being retained by the owner until it was assigned to the new owner as under the 1953 act, title in the surplus area was assigned directly to the state government (section 8). The payment schedule ranged from a minimum of "twelve times fair rent" to a maximum of Rs. 2,500 per hectare. The prices were plainly confiscatory, more so each year, although under Article 31 of the Constitution this by itself was not a legal fault. No provision was made for a family's land, once apportioned, to be regained by any of the original owners after the death of the father or son, in the traditional manner. Instead, this was precluded by the stipulation that no holder could subsequently add land above the ceiling amount. Land obtained by the state was to be allotted to "tenants, members of Scheduled Castes and Backward Classes and landless agricultural workers, of an area not exceeding two hectares of the first quality land or equivalent."

Since there was no provision that landowners and sons had to be within an actual undivided family, it appeared that this law required revenue officers to do what the previous act had carefully avoided: locate all the sons of the father, pool their holdings, and require the selection to be made therefrom (section 3b). That is, the definition of the family would be up to the government, not the people involved. Under customary law nothing gave a father the power to arbitrarily take the land of one son and give it to another or keep it for himself, either in an undivided or divided family. Nor was there anything in customary law that allowed a son to be deprived of an equal share in the family estate on the basis of age. Nor did customary law deny sons mutual rights in each other's holdings, or in the holdings of their father. Thus while the law was in a nominal sense more liberal than the previous act in one case—that of a man with more than one adult son in his house, but not recorded as a separate owner—in its general implications it posed the possibility of a radical and disruptive alteration of the mutual and interrelated kinship rights that up to then had been enforced by courts under customary law.

The new ceiling would have required one family in Shahidpur to reduce its holdings, had it not been that the family had enough adult sons at the time. In Panjab as a whole it was unclear how many families and how much land the law might have adversely affected, but since the law in effect would have reversed the trends toward larger operational holdings and toward greater specialization that were basic to the increasing productivity in the region, it was to be expected that sooner or later its provisions would run counter to the perceived economic and managerial interests of very large numbers of people, both farmers and nonfarmers.

Yet however many people saw their immediate practical interests threatened, the number who saw their moral framework threatened was bound to be far larger. By the unilateral power given the father (and the state), by the elimination of preemptive rights and rights of reversion, and by the denial of the birthright of minor sons, the act struck at the basis of the reciprocal cooperative relationship among men of a *parivār* and hence at the structure of the *parivār* itself. By implication, it also struck at the logical basis of the definitions of all the other kinship groups as well. Without the birthright of the son, neither father nor mother has security. Without preemptive and reversionary rights among brothers, there is no basis for cooperation among them. The entire structure of incentives and obligations that had framed the cooperation that produced the progress up to that point and that defined virtually everyone's self-conceptions, plans, and hopes, was dismissed.

The law was challenged in the Panjab High Court in 1973. The case, *Sucha Singh Bajwa* v. *The State of Panjab*, was decided in February 1974. The court ruled that as an effort to place ceilings on the amount of land a family could hold in order to spread productive resources, the law was indeed a measure of agrarian reform entitled to the protection of Article 31A, and thus immune from attack on the ground that it deprived people of property without compensation. Neither did it discriminate by sex, since it allowed female landowners to make the same selections as male landowners for their adult sons. But insofar as the act divided the land of the family without possibility of reversion, in effect eliminated the rights of preemption in ancestral property, and allowed land to be taken from some sons but not others, it was entitled to no such protection and it had to be amended to provide compensation (section 21).

With respect to the failure to preserve the family holding, the court held:

> The share of each member of the family has not been defined nor has any restriction been placed on the alienation of land by the members of the family so as to ensure its retention by the family. Such provision cannot be said to be in the interest or by way of agrarian reform, nay, it is the very negation thereof and cannot be upheld as valid or constitutional. (section 19)

The court also rejected the way the act required revenue officers to assemble the sons of a father into a single unit, regardless of the way they organized themselves, for the purpose of selecting the shares:

> The family, as defined in the Act, cannot be said to be an association or body of individuals unless it existed in that form on the appointed day and held any land in its name. By an artificial definition a "family" cannot be brought into existence retrospectively with reference to the appointed day and by a fiction deemed to hold land which was, in fact, not held by it but which was held individually by each of its members. (section 20)

Rather than amend the act, the state appealed the decision to the Supreme Court of India. At about the same time the act was placed on the Ninth Schedule, but no further efforts were made to enforce it. In 1978 this is where everyone I spoke to in Panjab thought matters rested. In fact, however, the Supreme Court had overturned the state decision in 1977, in a case originating in Maharashtra. The Supreme Court had not rejected the idea that the law required the creation of "an artificial family unit," but had simply held that such a device was protected by Article 31A and 31B (*D. G. Mahajan* v. *State of Maharashtra*, case 916; sections B, G).

Since I learned about the 1972 act rather late in the summer, there was not much time to discuss it in the village. I did, however, bring it up on two occasions with Indar Singh, his family, and some friends. Their first response was to point out what I have already noted: the new limit was still more than anyone in the village had, although it was closer. Their second response was to repeat the theme of *jātī* conflict that I had already heard several times before. It was explained that there was a clash of interests between Jats and Harijans. This act was seen as pro-Harijan and anti-Jat. It was taken as symptomatic of a consistent lack of appreciation of, or responsiveness to, farmers' needs on the part of the central government in general, and to some extent of an inclination on the part of the Congress party at the national level to appeal to a coalition of industrialists and Harijans against rural agriculturalists. Within this larger picture, the specific act was not of pressing concern. Since they felt that it might remain on the Supreme Court docket indefinitely, farmers were inclined to wait and see. No one, however, proposed that they would change their kinship concepts if the law was deemed constitutional.

The major Indian law reporters show no new case arising under the 1972 act until June 1979—a decision concerning the effect of consolidation before a relatively minor hearing officer (later overturned). In July the Panjab and Haryana High Courts heard a case wherein they ruled that land occupied by tenants could not be declared surplus (this decision moved a statewide political group representing farm laborers to call for a revision of the act). After that, there was nothing until 1980, when a fairly steady but not large series of cases began, revolving mainly around the relationship between the acts of 1972 and 1953. The 1972 act did not replace the former but held that proceedings under it should first be completed before the new proceedings were undertaken.

Evidently there had been little effort to enforce the 1972 act until the political alignment changed. In January 1979, the Janata party was defeated in national elections, and Indira Gandhi returned to power as Prime Minister. There were demands, publicized in the press, for the enforcement of this and similar acts in other states and for the creation of visiting commissions to look into such enforcement. In February 1980, the central government declared President's Rule in Panjab and eight other states whose governments were still dominated by parties other than Mrs. Gandhi's wing of the Congress. This dissolved the legislative assemblies and required new elections. In Panjab, the elections had the desired effect: the new government was supportive of Mrs. Gandhi's central group. Those formerly in power became the basis of the opposition movement that has continued to gain strength through the recent civil disturbances that have at times seemed to verge on being a drive for secession.

Despite the increased efforts of the central government, the act of 1972 has not been effectively enforced. According to press reports, as of January 1981, just 12,340 hectares had been declared surplus and 2,088 had been allotted. Evidently the previous land reform program had been more complete than the central government had believed. At the same time, the most frightening result of the enforcement of the act was avoided in the rules for its implementation. The revenue officers were not required to constitute families according to their records. Rather, the rules required each owner to make a declaration of his holdings and designate what he was to retain, including the additional fifth for each member of his immediate family (including minor sons and daughters) numbering more than five. Then the additional amounts could be given to adult sons who themselves had less than the limits. The fathers could not, under this procedure, arrogate the sons' land to themselves if the sons had already divided their holdings.

It would be wrong to suggest that the Land Tenures Act is the sole cause of the greatly increased tension between Panjabis and the central government, the rebirth of Sikh demands for regional autonomy, and the attendant civil disturbances that have erupted in Panjab since 1980. It is not, and it is not taken to be. It was an early cause of concern, but it was soon overshadowed by simpler and more dramatic issues which will be described briefly in the next chapter. Yet the new points at issue are considered by an apparently large majority of Panjabis to be shaped by the same combination of central governmental ignorance and callousness as the 1972 act. The farmers of Shahidpur, like the farmers in most of India (to judge from the pattern of similar disturbances in other states), know what the local system of production depends on. They can and do calculate their own self-interest, and they can and do formulate ideas for policies that would aid them as well as all around them. They are also willing to suffer some loss for the general good. Their objection to this law and the other measures they group with it is that they are bad both for the farmers and for the general good. The law aids a minority, and does that only superficially. It would curtail the trends that have brought progress by taking land from those who produce more and giving it to those who, in the nature of the case, would have to produce less and who probably have better opportunities elsewhere. At the same time it threatens the one institution on which absolutely all production depends.

One could debate the extent to which the act might miss its mark economically—the extent to which overall losses to productivity it might cause would offset the gains to those it is intended to aid or to Indians generally. But once one sees what the village systems are and appreciates kinship as a system of low information potential, there is hardly room to doubt that insofar as it attempts to remedy social inequities by attacking basic and longstanding kinship ideas, it is fundamentally misconceived.

I should add that by and large the group with whom I discussed the act had supported the Congress locally, and all had been quick to point out ways their present circumstances had benefited from other Congress ideas in the past. I should also add that villagers most frequently cited the reason for the massive rejection of the Congress in the state in the 1977 elections as being the government's attempt to meddle with kinship ideas—specifically the government's "arrogance" at trying to impose birth control. (The objection was not to birth control, only to the imposition.) I have no doubt they felt that village kinship ideas were supported by the vast majority in the region, and that if public opinion were again similarly offended, there would be a similarly firm and effective response at the ballot boxes.

In short, by 1978 basic kinship concepts had not changed. Despite a formidable subsequent legal and political effort, so far as I can determine they have still not changed, and they are not likely to do so.

CHAPTER 7

Religion

Most Panjabis, including most people in Shahidpur, are convinced that part of the credit for the Green Revolution must go to the Sikh religion. I agree, but at the same time I want to clearly reject current theories about the relationship between religion and economic behavior that are traceable to the tradition of Max Weber.

The relationship between religion and development is not causal, nor do some religions contain values that somehow promote rationality while other religions do not. The basic ideas of the Sikh religion have been constant for a very long time, and the Green Revolution is very new. There is no evidence whatever that Panjabis suddenly became more rational or progressive at the time of the Green Revolution or in any other way changed basic values that can be associated with the religion. It should be kept in mind that very similar economic changes have occurred in Haryana, where the predominant religious identification is Hindu.

Instead of causing the revolution, the Sikh religion has provided a conceptual vehicle through which individuals could argue for organizational and moral priorities that would coordinate individual and community interests. In this, what has been important has not been any particular directive force on behavior but rather the logical coherence of the ideas and their generality—their very lack of such specific import. The Green Revolution was created by people and not by mysterious social mechanisms, but to do it they used the ideas of the Sikh religion to reorganize and reorient the state government as well as to order and orient masses of the most minute personal interactions.

One reason for the importance of the Sikh religion in the Panjab context is that it provides an explicit and accepted ethical balance or counterpoint to the kinship ideas. The kinship ideas tie each person's interests to his or her family first of all and thereby make the family the main focus of economic activities and moral concerns. It has been recognized for centuries in the Sikh literature that if pressed to its logical extreme, this orientation leads to an ethic which is self-defeating, a point where one sacrifices the general social order in order to preserve one's family, although in the end no one's family will be safe in a society where everyone acts this way. Such a society is inherently weak, corrupt, and insecure. Accordingly, a major Sikh value has always been "martyrdom," self-sacrifice, advocated as the opposite of "selfishness." "Self" in this context includes family. So self-sacrifice does not mean individual self-destruction but rather sacrificing one's immediate personal and family interests in favor of the interests of the general community.

The idea of martyrdom has obtained its present prominence through a long series of historical developments. Judging from the apparent age of little monuments to martyrs (*shahīds*) throughout the villages of Panjab (there are over fifteen in Shahidpur), the idea has been used at the village level for at least three centuries. At higher levels of organization, it was used by the last Sikh gurus to define their supporters, as will be described, and has been associated with the martial traditions of the Sikhs ever since. Its modern nuances of meaning go back to the beginning of this century, when it was used to articulate calls to join the general movement for Indian independence and to join groups more particularly aimed at creating a "homeland" for Sikhs. After India gained independence in 1947, the call for a homeland continued to be sounded, and the idea of "martyrdom" continued to be invoked in arguing for its support. Ultimately this movement was transformed in the successful drive to separate Panjab and Haryana. When the new state was finally formed, the policies and programs that had been used to argue for the division and the interests that were mobilized by these arguments naturally became the policies and interests of the state government itself. The progress since then has been at least partly because of this.

At every point in recent history, the broadly advertised and accepted interpretations of basic religious ideas were echoed in local settings, and local uses in turn showed support for and added depth and meaning to the large-scale ideological discussions. Religious gatherings in the village and in the many major Sikh gurudwaras throughout the state provide the settings for a active circulation of people and ideas. Speakers come to villages; villagers travel to speak and to hear others; and what is heard and said is a constant topic of village discussion everywhere.

As the ideas of the religion were given interpretations that people continued to agree upon and consider important, the organization that produced this consensus was inevitably supported as well. At every point in history, including the present, it has been impossible to separate the impact of the religious ideas from the impact of the organization that these ideas rationalized and that has developed and promulgated the ideas and their applications.

Partly in connection with the independence movement, this century has seen the development of a kind of renaissance of Sikh literary and philosophical activity. In 1885 Max Arthur MacAuliffe, a native of Ireland posted in Panjab as a magistrate, undertook to collect and set down the history and basic ideas of what he then perceived to be a noble religion on the brink of extinction. Sikh literary activity has been increasing ever since.

The basic teachings of the religion have always been precisely formulated and form a tightly integrated system. At the present time, there are specifically designated authoritative works and public ceremonies that represent them. There is a system of formal education culminating in degrees at the baccalaureate and graduate levels. Finally, as one aspect of the resurgence of the religious community in this century, there is a legally established system of management of the major Sikh temples (designated historic shrines) that is concerned with promulgating the religious ideas and explaining their relevance to ongoing events on a daily basis. This system of management is headed by officials elected by universal suffrage among Sikhs (with no test of the status of a voter as a Sikh except self-declaration). The main administrative body of this system, the Shromani Gurudwara Prabandak Committee, or the SGPC (literally "Head Gurudwara Management Committee"), also supports schools, oversees the publication of authentic copies of traditional religious texts, and supports publication of current historical and linguistic scholarship and exegesis. The effective headquarters of the system is the Golden Temple at Amritsar, one of the most important historical gurudwaras and the site of the Akāl Tākht (literally "the immortal throne") from which policies are traditionally announced.

THE CENTRAL IDEAS

It is often said by Western religious scholars that Indian religions in general are nonhistorical. This is about half right. The main religious writings in the Indian traditions do not purport to describe a sequence of events since a Creation, but rather describe God as a timeless basis of knowledge. Yet accompanying these main writings is a very substantial body of historical lore, which constitutes the main exegetical material. The histories of the gurus explain the main ideas in the religious poetry and describe the institutions established by the gurus. Institutions, ideas, and usages not grounded in these histories are considered customary or have no standing in the religion proper at all. In this sense, for example, the SGPC is not an institution of the Sikh religion, since it is established *only* in law. I will follow this exegetical usage here, providing the general ideas and explaining them historically.

A second point often made in the West is that Sikhism is a syncretic blend of Hindu and Muslim ideas, perhaps predominantly the latter (see Hutchison 1981:186; Noss 1980:220ff.). The same theme was sounded in Kushwant Singh's history of the Sikhs (1963), to the dismay of many Sikh scholars. It is not correct. Sikh ideas accommodate a number of important Muslim themes, most notably militancy and the absolute equality of all those within the community. But they do so on the basis of premises that are thoroughly Hindu.

The Sikh ideas are grounded in the tradition of Vedantic scholarship called bhakti, which emphasizes the importance of personal (rather than communal or conventional) devotion. Vedant, the religious philosophy of the Upanishads, is the double idea that religion is an attempt by "the disciple" to understand God, and that God is the basis of all understanding. This does not mean that God is an entity of some sort, among whose attributes or activities is being the basis of knowledge, as in the Semitic traditions. It means that God is nothing but the basis of all knowledge. Whatever is this basis, that will be taken as God. Accordingly, God is defined as "formless," "placeless," and as something other than "any moving thing in the moving world." Given this, God is described as in everybody, and yet as more than "just the self" (I take all these phrases from the Iśa Upanishad). Man is not made in God's image, nor God in man's. They are not alike but rather fundamentally unlike. The relationship between God and the disciple is often likened to the relationship between a fish and a pond. The worshiper is the fish; God the pond. The fish has a birth, life, and death. It has shape while it lives and while it lives it draws all its substance and power from the pond. Anything the fish does and anything it knows, including itself, is through the pond. The fish comes from the pond, and will upon death dissolve back into the pond. Yet the water of the pond itself is shapeless and neither living nor active. It has no mind and no power of movement. Similarly, God supports and sustains the disciple, is within the disciple and around him, and certainly other than the disciple though not apart from him. The main collection of Sikh religious writings, the Guru Granth Sahib, consists of poems which develop similes such as this. Each in its own way illustrates the nature of God. Most also spell out the ethical implications of this position.

Within the broad Vedantic tradition, bhakti emphasized that each person has to arrive at his own understanding of God in his own way. From this follows the importance of a personal teacher, the *guru*. The founding teacher of Sikh tradition was Guru Nanak (1469–1543). Two of his first pronouncements indicate his specific application of the more general ideas. The first of these was "There is no Hindu, there is no Muslim." Since, at the time, there was bitter warfare among the two groups, the remark naturally required explaining. Eventually the explanation was given: "Only God is the true guru." The statement reverberates with much of the sense and sentiment of the Muslim evocation "There is no God but the one God," but it is at the same time in spirit and substance both pure Vedant and very sophisticated bhakti. The idea is that God as such is equally separate from all specific systems of teaching, all "gurus" (thus quietly denying the authority of prophecy, which is fundamental to Islam as to the Judeo-Christian traditions). Accordingly, no one system could claim to be the one true system, to uniquely contain or embody God. Each was but one attempt at understanding. In that same sense, there was no Hindu who was something with respect to God radically different from a Muslim and no Muslim who was something radically different from a Hindu. There was no need to choose to be one rather than the other and indeed in a fundamental sense a person could not be one rather than the other. One was a disciple or one was not; one tried to understand God or one did not.

Guru Nanak's advocacy of absolute religious tolerance is the foundation of the Sikh system of ethics. Nanak's poems do not lay out a series of substantive commandments, but rather describe an attitude toward God and toward oneself. They contain frequent criticisms of claims of special religious privilege or knowledge, of exotic forms of worship thoughtlessly pursued out of habit or for the sake of claiming a privileged social position, of penance, and of the life of the mendicant. Instead, the poems argue that any act in ordinary life can be a form of worship if done with one's thoughts on God. Thus the life of the true disciple is not to be lived cut off from the world or in performing unproductive acts, but rather within the world carrying out productively the responsibilities and obligations of daily life. The term "Sikh," from the verb *sikhNā*, "to learn," means disciple in precisely this sense. It follows as one particular application of this general orientation that for a Sikh there would be no point whatever in attempting to force the conversion of anyone to his religion, or to attempt to threaten or otherwise induce any Sikh to avoid displaying an interest in any other religious system. The ideas were consistent, complete, and powerful.

Logically it may seem that Nanak's position should lead to a kind of pacifism and there are some Sikhs, especially among those who are called

Nānakpanthīs, who believe it does. But history after Nanak forced further developments and these eventually produced the Sikh idea of martyrdom.

Initially, Nanak and other scholars in the bhakti movement, most notably Kabir, attracted considerable support from both Hindu and Muslim rulers. Nanak's descendants and other Sikhs were often given small principalities to form buffer states between Hindu and Muslim areas which led eventually to Sikhs gaining significant strength in the most pivotal and insecure areas of the region. With the fifth guru, Arjun Dev, this itself came to be perceived as a threat by neighboring powers, especially Muslim powers, and tolerance turned to attempts to assert dominance. The Sikh response in turn was militancy, in the form of several interrelated institutions. First, the religious training of disciples was made to include military exercises. Second, the guru instituted the communal kitchen, or *langar*. Wherever he taught, food was made available to those who came to learn. The food had the practical purpose of keeping together a viable bodyguard, but it had a symbolic significance as well. It represented in a material form the sustaining power of the teachings and ultimately of God. Finally, the guru himself, in addition to being a religious teacher of the first order (writing poetry of extraordinary beauty and power), became a military leader as well.

Militancy as such is not a Sikh virtue. The Sikh virtue that unites these institutions and produces the effect of militancy is self-defense. When a person identifies his interests with those of his community and sacrifices himself to defend both, it is martyrdom. The basic literature of Vedant contains strong admonitions against "violence" and in favor of "nonviolence" or "nonattachment," but there is nothing in favor of suicide. The basic Sikh idea of self-defense and its connection to martyrdom is given in the history of the death of Guru Arjun. Putting it starkly, he was taken by force and given the choice of conversion or death. He refused and was tortured to death. He died without complaint, explaining his views to the end. The underlying idea is that conversion would not really have saved his life at all. If a person has an understanding of God which he considers to also be an understanding of the basis of his own existence, an understanding that sustains him as food does and as the water in the pond sustains the fish, and if that person is then faced with the demand that he abandon that understanding, the demand would be the same as requiring him to give up life itself. In the Sikh view, to accept such a demand is the same as committing suicide. It is better to resist, if possible to fight. The justification for fighting will end at the point where one's defense, which means the defense of truths one depends on, is secure.

Guru Arjun's ideas were developed by subsequent gurus up to the tenth and last, Guru Gobind Singh, who gave the religious community its modern form. He replaced the line of individual gurus with a self-perpetuating organization by a double invention. First, he completed the collection of the writings of the gurus and similar thinkers that had been begun by Arjun Dev. This became the Guru Granth Sahib. Second, on Baisakhi Day in 1701 he inaugurated a ritual of initiation and stipulated that henceforth wherever five initiated Sikhs assembled and acted in the presence of the Granth Sahib, they would be acting as the guru. After Guru Gobind Singh set up the ritual and initiated the first five Sikhs, his next act was to ask them to admit him to the community, the *khālsa*.

The explicit aim of these reforms was to reduce the vulnerability of the religion to disputes over succession and to military suppression. Disputes over succession had already threatened to take ideological form and obscure the main themes that all agreed on. With the book there would be clarity of thought, and with the community of Sikhs there would be no one person who could be sought out and killed but a chain of guardians who could disperse and continue to grow.

Gobind Singh's ritual of initiation provides the main symbols used by modern leaders to convey the religious ideas. The first five Sikhs had responded to the guru's call for five who would "give their heads" to the guru as a sacrifice. It must be understood, to appreciate the story, that the idea of human sacrifice in this context is utterly abhorrent. Since the whole purpose of religion is to inform, sustain, and enrich life, the logical reaction to any who makes this kind of request is that they must be mad. In addition, a person who allowed himself to be killed was necessarily abandoning his obligations to family and kin. Those who volunteered, each facing clearer and clearer evidence that the guru meant the request literally, are understood as demonstrating confidence that the guru was not mad and that, despite appearances, he was not asking for something inconsistent with the teachings of the previous gurus. They therefore showed the mixture of confidence and courage that the guru was seeking. The ritual itself links this devotion to the community over family to readiness to fight for the community; it is said and enacted that as the initiate severs his relations to his family symbolically now, so he cannot call upon his family obligations later to refuse a call to defend the guru and his community. The main action of the ritual revolves around the initiates accepting sweetened water (*amrit*) from the hands of others previously intitiated, whose castes are unknown, and those accepting it are told to call each other *bhāi* henceforth. This reinforces the rejection of caste while also symbolizing the sweetness of the truth and the idea that it sustains life—as food itself. That is, the patrimony of kinship is replaced by the patrimony of the guru's teaching. There is no Sikh priesthood. Any Sikh may read the Guru Granth Sahib and teach.

CEREMONIES

Such general ideas as the basic concepts of God, the guru, and the Sikh obviously cannot automatically lead public opinion in any particular direction. They must be applied actively and carefully and great efforts must be made to gain consensus on such applications. The ceremonies provide the framework for these efforts. Strictly speaking, apart from initiation there is only one Sikh ceremony: the reading of the Guru Granth Sahib. Ceremonies that may seem distinct, like those for birth, marriage, and death, are just different customary elaborations on the one pattern and have no historically established special status. Consistent with the gurus' criticism of ritual for its own sake, there are no "sacraments" in the Christian sense.

The Guru Granth Sahib may be read in its entirety or in part. In what appears as the ordinary daily service, just parts are read. Specifically, after a selection of poems that includes the first writing of Guru Nanak and the *Āsa dī Wār*, as well as other pieces which taken together include at least one poem from each of the gurus who contributed to the Granth Sahib, one poem is selected by letting the book fall open to a page by chance.

Reading the Granth in its entirety may be done continuously, which takes three days and two nights, or with interruptions, which takes about a week. The continuous reading generally involves three readers' taking turns; the interrupted reading is usually done by one person. The readers are almost always hired from outside the village for the occasion. Continuous readings are generally sponsored by private individuals to celebrate some good fortune or to mark the beginning of an enterprise.

Whatever the type of reading, it falls within a generally standard framework. First, the reading is always done in the presence of at least five Sikhs. Second, food is always served in at least two forms: *krāprashād* which is served in a ceremony after the reading and before the Granth is closed, and the *langar* which is served concurrently with the reading. Third, heeding the injunctions and examples of several of the gurus, the reading is always followed by explanations or "instruction," called *sikhīā*. All of these features work together in making the ceremonies a major medium of public debate. The poems state and illustrate the principles; the food permits a larger group to gather and emphasizes that the principles apply as a basis of life in this world; and the *sikhīā* provides the opportunity for people who are important in the village and region to come forward and present what will count as a contemporary example of the abstract principles described in the poetry—those actions of the day they consider to illustrate the truths of the gurus or events and actions in Sikh history, good or bad.

The bottom tier of the system of religious communication that knits the

Panjab together consists of a large number of "preachers" who travel from village to village singing or speaking in these ceremonies. By knowing who they are, what they are saying, what the reactions were to what they said, and who invited them to following ceremonies, any knowledgeable person can form an extremely precise opinion of the various opposing positions on current concerns and the support for each. Publicly sponsored readings, such as that of the annual fair in Shahidpur to honor the martyrdom of the sons of Guru Gobind Singh, where the controlling local committee selects the singers and speakers for the *sikhīā* generally have a covert political coloration. People who support the committee's orientation come to listen; those who oppose it stay away. At larger gurudwaras, on major holidays, several political groups may sponsor *sikhīās* concurrently in different parts of the building or grounds. Each will have its own supporters and each group of supporters will boycott the others (but will note who is there and how many).

In other types of gatherings, such as overtly political meetings or in speaking directly to the press, the same style of argument is used as in the *sikhīā*, although in such cases the political content is often more explicit. At the same time, the general demeanor expected of a Sikh teacher is affected by politicians (see Leaf 1972:157ff. for a fuller description). It is no accident that one of the major religious figures in this period was titled "Master" ("schoolmaster") and the other was "Sant" ("saint"). In fact, in style of life and manner of speaking it is impossible to draw a line between full-time politicians and the preachers, and such preaching is generally regarded as the first step in a political career.

In Panjab in 1978 and 1965, there was no question that if one wanted to know what policy issues were being discussed, what the positions were, and how much support each position had, by far the best way to learn this was to make the rounds of major gurudwaras and a sample of local festivals and listen to *sikhīās*.

REGIONAL CHANGES IN CONSENSUS: 1965–1978

On January 17, 1965, there was an election for the officers of the Shromani Gurudwara Prabandak Committee. Shahidpur was a polling site for this election as well as for a "local" election for the presidency of one of the gurudwaras officially designated as within the SGPC purview, at Chamkaur. In both elections there were over a dozen candidates, but all or almost all were pledged to support just one or the other of two major contending national (communitywide) leaders, Master Tara Singh or Sant Fateh Singh.

Accordingly, they were described as supporting, or being "of," either the Master Tara Singh party or the Sant Fateh Singh party. Whichever group obtained the majority on the SGPC would elect its man or his nominee as chairman. He would thereby become the major public spokesman for the Sikh community in general, to the extent that any single person might claim such a role.

Each candidate was identified with a definite platform, in sharp disagreement with the other. The disagreements primarily concerned economic policy and the basis upon which Panjab should be divided to create a *panjābī sūbah* ("Panjabi state").

The history of the debate over the Panjabi Subah in the January election was virtually coextensive with the Sikh renaissance and reform movements already noted. One reason why Sikh demands for an autonomous region were ignored prior to 1947 was that there was no geographical area where Sikhs were more than a large minority. The partition of pre-1947 Panjab between Pakistan and India had the effect of dividing the Sikh population almost exactly in half, with the result that they suffered proportionally by far the greatest losses in the resulting migrations, confusion, and violence as those Sikhs who were on the Pakistan side of the border fled to India. This very migration, however, created for the first time a substantial region where Sikhs were numerically preponderant: the central zone of the new Panjab (India) State.

The Indian Constitution outlawed "communal" organizations and demands and ended the British practice of structuring legislatures by religious, caste, and ethnic blocks of seats (except where necessary to safeguard minimal representation for certain minorities). At the same time, however, the Constitution not only permitted a focus on languages for defining group and regional interests, but virtually mandated it.

The mandate lay in the constitutional declaration that Hindi would become the national language of India effective January 26, 1965. As the date approached and then passed, tension increased. The provision had strong support in Hindi-speaking regions but increasingly strong opposition elsewhere. At the national level tension was allayed when Prime Minister Lal Bahadur Shastri announced on February 11 that English would be given "associate" status with Hindi and that each state would be able to conduct internal business in the language of its choice. This met some of the demands of the Dravidian language areas in the south, but it only sharpened the debate within Panjab.

In 1965 both Hindi and Panjabi were official languages in the state, but in the government, where the importance of "national unity" was being invoked, there was much discussion about increasing the emphasis on Hindi. The Chief Minister at the time was Ram Krishan, both non-Sikh and a Hindi speaker. Sikhs were extremely concerned with possible developments. It was explained to me in and around the village that everyone knew that teaching a language required a literature. It was generally considered that the main Hindi literature was the great classics associated with Hindu religious ideas, such as the Ramayana and Bhagavad Gita. Panjabi was especially associated with the Guru Granth Sahib. English was not associated with any particular religious literature but rather with government and commerce. It followed that if more emphasis were placed on Hindi, especially in the schools there would have to be more time spent reading Hindu religious works and in effect teaching Hindu religious ideas. English would suffer and so, necessarily, would Panjabi as well as the Sikh literature and the viability of Sikh ethical and religious ideas. Every rural Sikh I spoke with was concerned about this and resented this possible use of governmental power. They also assured me that there was no disagreement on the point anywhere and most saw it as a reason for seeking a separate Panjabi-speaking state. This is why the differences of opinion in the election concerned only the *basis* for the creation of such a state.

Master Tara Singh was associated with the view that the new Panjab should be a Sikh majority state, on the ground that only such a state would secure the Sikh community and religion from political persecution. His argument rested on the legitimacy of self-defense and insisted that the new state's boundaries should include whatever areas had a majority of people who declared themselves Sikh by religion.

Sant Fateh Singh claimed that Master Tara Singh's plan endorsed religious discrimination and was inconsistent with the Sikh view of the equality of all religions and the tradition of tolerance. He argued that the boundaries should be drawn strictly along linguistic lines and that there should be no attempt either to have a Sikh-dominated government or a government run on specifically Sikh "theocratic" lines. Areas with a majority of Panjabi speakers who were not Sikh should still be included. Areas with a majority of Sikhs but not a majority of Panjabi speakers should be left out.

The boundary implications of the two policies were genuinely different, as were the implications for the subsequent system of government. There were Panjabi-speaking areas to the northeast and to the southwest of central Panjab—mainly the districts of Kangra and Sindh, respectively—that were widely recognized as being Hindu, or at least not particularly Sikh, by religion. Thus it was quite possible that a Panjabi majority state might not be a Sikh majority state.

In the realm of agricultural policy, there was a general feeling that the central and state governments in 1965 favored the Hindu majority areas in the southeastern part of the state at the expense of the central Panjab where Sikhs were predominant.

The agricultural system of Shahidpur is fairly typical of the central area as a whole, as noted. In the predominantly Hindu areas of the southeastern portion of the state, there were fewer and smaller towns, lower rural population densities, lower rates of geographical mobility and, on the farms, lower cattle densities, higher proportions of acreage in drought-resistant crops such as chickpeas and millet, and heavy reliance on irrigation from canals alone or rainfall land alone. This is an area of low and irregular rainfall and a deep water table.

At the time, government development efforts concentrated on extending the canals and digging very large tube wells from which many farmers would buy water. The aim of such projects was to provide the maximum area of assured subsistence, one good crop a year. This meant that the water was to be dispersed as widely as possible in the southernmost regions.

The central Panjab farmers objected to the central government's policy for several reasons. First and most obvious, it promised them no benefits. Farmers in the central area were already farming above subsistence levels. Second, it actually caused positive injury. Canals in raised banks, which all Panjab canals are, necessarily disrupt the normal paths of drainage. In the central area when rainfall was even moderately heavy (at least a few times in most years), they acted as high dams, blocking runoff and causing floods. In addition, in some areas percolation from the canals brought the underground water all the way to the surface, causing waterlogging. In the winter of 1965 large areas of central Panjab, especially in the important Amritsar district, were lost to the usual crops. There was much alarm and government reclamation efforts were considered futile and halfhearted. (Eventually these areas were provided with drainage channels, planted in rice, and made quite productive.) Since the southern districts had so little rainfall and such low water tables, the canals did not bring them these same problems.

The farmers of the central region saw no good purpose in a policy that extended a form of agriculture that provided no marketable surplus for the country as a whole and rendered useless some of the most productive areas. In the nature of the case, it was natural for many of the farmers to believe that the policies were not so much a result of simple incompetence or insensitivity as a deliberate attempt to weaken and injure Sikhs and the Sikh areas as such.

Beyond the water policies, farmers in the central areas felt themselves further injured by certain economic measures the central government had put in place in response to the poor national harvests in the summers of 1964 and 1965. These included the institution of rationing in the cities and the establishment of "food zones" over the rural areas. The boundaries of the zones cut Panjab off from important markets in Delhi and Uttar Pradesh. Trade in the major foodstuffs was permitted as usual within such zones but prohibited between them. Instead, the government bought foodstuffs in each zone, moved them across the borders, and marketed them through "fair price shops." This depressed prices within the surplus areas below what would otherwise have been the case and, since the fair price shops were located in cities, not villages, many rural people saw the entire plan as an attempt to subsidize the Congress party's political base in the cities at rural expense. They resented it and said that it, again, would simply deny them the resources they needed to further develop their productive capacities. Again there were many to suggest that this denial was not based on purely economic motives but reflected religious or cultural hostility as well.

Both the major parties in the 1965 SGPC election agreed in opposing these central measures and argued that what was needed was a set of agricultural policies to support and encourage small-scale independent farmers and provide incentives to commercial production. The main disagreement was over the role of land reform, especially limitations on the size of holdings, in such a policy.

The Master Tara Singh party was considered to be allied with the Congress party, which at the time was considered to favor the larger landholders by opposing lowering the holding limit and redistributing land to landless persons, be they tenants or Harijans. The Congress was also, of course, the party in power and the author of the policies regarding canals and food rationing. Master Tara Singh denied the link to the Congress and his supporters claimed that this "neutralized" the issue.

The Sant Fateh Singh party did not agree. In its turn, it was considered to be allied to the Communist party and to be in favor of lower holding limits and land redistribution but against large-scale government schemes such as the canal projects.

The religious debate relevant to the land reform issue concerned the interpretation to be given to "self-sacrifice." The supporters of Master Tara Singh held that this was a purely individual matter and had nothing to do with land reform or legal compulsion generally. In their view, the relevant religious principle was the importance of the individual, which should logically lead to policies securing individual rights in the property they had worked to build up. The supporters of Sant Fateh Singh held that the concept of self-sacrifice did apply and that those with too much land should be willing to give some of it to those without enough for the good of community as a whole.

When the Sant Fateh Singh party won the SGPC election by a comfortable margin, its interpretations of all these issues came to be the ones used in religious events at major Sikh shrines and in official policies of the gurudwara organization, both because the majority of the committee controlled the selection of speakers and because the minority kept silent. It is not considered respectful to carry political conflict (as such) into religious activities. At the same time and equally important, the Sant's view become the dominant position of the Shromani Akali Dal. The Dal was and is a political party made up primarily of Sikhs that draws most of its leadership from the religious hierarchy, although strictly speaking it is legally and financially separate from the SGPC. In practice, however, the Dal acts as the political arm of the Sikh religious organization and as such contends with other political organizations, such as the Congress, in the regular elections (rather than the gurudwara elections). The SGPC itself cannot enter elections since it is legally a management trust for the gurudwaras and only that.

For the general elections of 1966, the Akali Dal published a manifesto covering all the points just described, among others, and illustrating the way the issues were brought together within the religious frame of reference. It included fourteen major points, seven of which were directly related to the central region's distinctive pattern of agriculture. For example, the Dal:

> favours forms of state and measures, such as arouse the creative potential of the individual endeavour and enterprise harnessed for the common good within the restraining framework of a good and honest government, with a view to destroying and eradicating the stagnation and fatalism entrapping Indian society;
>
> favours self-owned and self-cultivated viable agricultural farms in Punjab run with generous state aid on modern lines, such as must form the basis of a healthy, egalitarian, proud, freedom-loving and progressive society capable of successfully meeting crises and resisting oppression from whatever quarter, that of the state or other form of tyranny;
>
> favours agriculture-based industrialization of the Sikh homeland and farm-based dairying and poultry-keeping, coupled with aforestation of its sub-mountain areas and encouragement of other small scale industries feasible for these areas, such as textiles, woolen and machine tool industries;
>
> favours growth of natural leadership and cooperative initiative, free from the fetters of needless state controls and interference.

The Panjabi Subah was linked to these policies as the best guarantee of a state government that would carry them out and the only way to oppose effectively "a section of political Hindus [who] . . . have a firm plan for degrading and liquidating the Sikhs from the mainstream of a free India" (section 8).

After the election, Sant Fateh Singh intensified his drive for the Panjabi

Subah defined along linguistic lines. He soon announced that if a *sūbah* were not created, he would set a date to begin a fast to the death.

Westerners often say that Indian hunger strikes and threats of self-destruction show a low regard for human life. Exactly the opposite is the case. The power of the gesture depends precisely on the underlying assumption that nothing is more important than life and that therefore the only thing that would lead a person to risk his life this way is his expectation of a more certain death if he does not. Given this, the significance comes directly from the idea that religion is the basis of one's life. For a Sikh, as such, to undertake a hunger strike in protest against a policy is to announce symbolically that he believes that the policy will destroy his religious patrimony.

If the striker is subsequently allowed to die, this would support his claims. It would not alone serve as an outright proof, since one can always be sincere but mistaken. But if the striker argued his case convincingly, then for those with confidence in his judgment it would follow that the only logical and self-respecting course was to risk death to defend the principles until the fight was either won or lost. Better to rise and fight at once and together than be worn down and destroyed piecemeal.

For the person undertaking the hunger strike, the risks are extreme. If he does not win his point, he is almost sure to be destroyed either politically or physically. If he overestimates his support, he might die and still accomplish nothing. Worse, he might die and be considered a lunatic. If he sees no victory in sight and abandons the fast, those who accepted his view of the issues will consider that he has betrayed them through lack of courage.

No one had better reason than Sant Fateh Singh to know the logic of the course he proposed. His own rise to prominence began in 1961 when Master Tara Singh failed to carry out a hunger strike over an autonomous Sikh state after a series of such strikes that ended in obtaining concessions which were not subsequently honored. The strike occurred in the main Sikh gurudwara, in Amritsar. After forty-three days, obtaining no results, he gave up. Despite being one of the founders of the modern Sikh social renaissance and a leader in the independence and autonomy movements virtually since the first decade of the century, he was tried by a Sikh religious court and "found guilty of proving false to his oath and to the hallowed tradition of Sikh martyrdom" (Kushwant Singh 1963, vol. 2:300). He was sentenced to clean the shoes of the congregation for five days. This he did as assigned and later continued to attract a following to a degree, but he never regained his previous standing or undertook another hunger strike.

It was clear from the network of endorsements of Sant Fateh Singh's initial announcement that his assessment of the situation and its remedy was widely shared. In addition, his cause was aided by the war that then erupted between India and Pakistan.

Border clashes between Indian and Pakistani troops had been occurring all summer in 1965. On August 15 Pakistan attacked in battalion strength in Jammu. India replied with a series of actions culminating in sealing off the entry routes in the Haji Pir pass on August 28. Pakistan then launched a massive armored attack across the border in Jammu, and this began a series of armored thrusts and flanking attempts across the plains that quickly spread southward down the entire border of Panjab and into Rajasthan. There were blackouts in the cities and towns and fears of spearheads breaking through toward New Delhi itself.

The combat underlined the strategic importance of the Sikh area and of the Sikh components in the Indian army. Sant Fateh Singh met with Prime Minister Shastri on August 7 and 8 and expressed his views strongly, then returned to report the results of the discussion to the official working committee of the Akali Dal. The Panjab papers were full of comments and reactions that made it clear that while there was some strong opposition, notably in the state government, the Sant had overwhelming and growing support for his demand in the rural areas. The Dal committee decided he should begin a fast (meaning taking only water) on August 16. If the demands were not granted, the fast would end in self-immolation at 9 A.M. on September 25.

In the first week of September, the Sant and SGPC representatives had more talks with the government in Delhi. The war was heating up; there were air raids in several Panjab cities, movements of refugees, and reports in the papers of Pakistani paratroopers being dropped on suicide missions, to disrupt communications and create terror (none did). The Sant was prevailed upon by the government to break off the fast, but this did not stop the agitation so much as push it in a slightly different direction. It had become clear that the Panjab could not be divided to please the central region alone, and the case for division was greatly aided by the formation of a parallel movement for a state of Haryana, based on Hindi, in the southern districts. The Sant encouraged this and there was much public discussion of possible boundaries. Toward the end of September 1965, after the fighting ended but before a peace treaty was signed, the government announced formation of formal bodies to consider the question. On March 18, after Indira Gandhi had become Prime Minister, a parliamentary commission recommended trifurcation of the state. The enabling act was passed in September 1966, establishing Presidential Rule while the new governments were formed. Formation of the new governments included the division of civil service personnel along linguistic lines, with the result that each state gained a government whose officials were more knowledgeable of, and sympathetic with, the needs of its rural people. At the same time, each government took on a religious (and cultural) coloration more strictly that of its region: the government of Panjab took on a more Sikh style, that of Haryana more Hindu. The new styles naturally were considered to signify the new responsiveness in both cases. Presidential Rule ended at midnight, November 1, 1966, and the new governments took over (see Gulati 1974, for a fuller account).

The new Panjab government immediately began to implement the policies that the Akali Dal had argued for, with the effects described in Chapters 3 through 5. This quite naturally reflected favorably on the religious organization and strengthened its support, but it did not arrest its evolution. New difficulties arose and the response to them led to further changes.

The new state government was dominated by the Congress party, not the Akali Dal. It included three prominent Congress supporters who had been under criminal indictment, which was offensive. Chandigarh had been made a separate federal district under central government control to serve as the capital of all three new states. This was itself a blow and led to charges that there were so many "linkages" between the new governments of the three states that the independence of the new Panjab state was undermined. Sant Fateh Singh promptly began a hunger strike for Chandigarh's inclusion into Panjab alone. Here, however, the case for the survival of the Sikh patrimony being at stake was much less plausible, while at the same time the government was much less free to concede the demand. In the face of virtually certain failure, Sant Fateh Singh, like Tara Singh before him, gave in to pleas of his relatives and secretly broke the fast. It was discovered and his credibility too was destroyed.

Despite his personal fall, the secularism that Sant Fateh Singh had articulated was consistently carried into the party politics of the new state. Not only was the initial domination by the Congress consistent with this, but so was the way in which the Akalis formed the opposition. They quickly allied themselves with the Jan Sangh and the Communist party. The Jan Sangh is a militantly conservative and militantly Hindu party which draws primarily from urban professionals and merchants. Up to the date of the division of the state, the party had opposed the Akali demands.

The Akali coalition came to dominate the state in the 1969 by elections In 1971 the Congress split and a section, with the Communists, formed part of a new dominant coalition. Still later, in the 1977 election that saw the defeat of Indira Gandhi nationally, the Communists were virtually wiped out locally, while the Congress formed the opposition minority and the dominant group was again a coalition centering on the Akali Dal and the Jan Sangh, going by the name of the Janata party from a national point of view. This was the state of affairs in 1978.

By 1978 an interpretation of the religious ideals of toleration and neutrality had become part of the structure of the state government. The same was true of the agricultural policies first argued for with the ideas of individuality, independence, and self-sacrifice. In this sense, religious concepts that were ambiguous in 1965 had become settled and clear by 1978. At the same time, however, the Sikh religious organization that had originated these policies did not become the government. It continued to be in a position to criticize and did so frequently and effectively.

RELIGIOUS CHANGE IN SHAHIDPUR

A discussion of religious change in Shahidpur itself must take into consideration the purely local shifts in the use of religious ideas as well as the discussions that were part of the state and national debates. In the larger debates, most villagers had supported Sant Fateh Singh's position in 1965 and continued to adhere to the same basic ideas in 1978.

The most obvious of the more local changes was not a qualitative shift in opinion so much as a change in intensity and interest. The first indication of this was the physical improvement of the village gurudwara itself, as noted.

The remodeling was described as having begun in 1967, with a private gift of Rs. 1,800 from a landowner of the village by the name of Giani Gurmail Singh. (The title *giānī* indicates the equivalent of a baccalaureate degree in Sikh literature and philosophy.) This amount bricked the courtyard and provided new doors. In 1969 the new dharmsala was built at the end of the courtyard opposite the main building, doubling the total floor space of the complex. It was paid for by a collection organized by the panchayat. The assessment was Rs. 10 per house, Rs. 1 per *bīghā* and Rs. 1 per "turban" for those who had no land. In addition, the panchayat sold two *bīghās* of common land, near the *abādī*, to those who had worked in the village as partners of farmers, to be used for work areas, for Rs. 5,000 per *bīghā*. The government also gave Rs. 110 (the unbalanced number signified that the relationship was to be ongoing, not stable and settled). This is the description of the remodeling as it was given to me by Indar Singh and it is appropriate to the ideas of the religion that no *jātīs* are mentioned for any of those involved. My notes do not record the point and my memory is uncertain, but I believe Indar Singh also said that the assessments were not enforceable. Some gave and some did not. The total amount raised was Rs. 5,000. The building was electrified in 1975 by another subscription.

As noted, the new *granthī* for the gurudwara had come only in 1977. The interest he showed and the level of his daily activity in pursuit of his basic duty of reading out the Guru Granth Sahib were far greater than those of his predecessor in 1965. Since my room at that time had been on the second floor of the gurudwara and connected to the main interior area by two windows, it was hardly possible to fail to notice any activity. Yet all I saw other than the large annual religious fair commemorating the flight of Guru Gobind Singh from Moghul persecution (which was not organized by the *granthī*), was a *sangrānd* at the beginning of each month. This was usually attended by just a few women and children, not working men. A *sangrānd* is basically an ordinary daily service with *sikhīā*, a random selection from the Granth Sahib, and distribution of *krāprashād*. The *granthī* would also sometimes take part in individually sponsored readings in village houses.

In 1978, in addition to the *Āsa dī Wār* each morning, the *granthī* conducted a quite serious discussion group every evening. This too was broadcast over the public address system. He also made sure that the Granth was on display in the traditional manner and available for those who would come to pay their respects virtually at any time.

Accompanying these changes in the intensity and frequency of formal religious activity was a subtle series of villagewide changes in dress and manner. It seemed to me as I went about the village that more of the villagers were adopting more of what had been considered the "typical" (as opposed to religiously recommended) dress of Sikh—the steel bangle on the right wrist, a certain way of tying a turban, and a full beard tied up neatly, along with some preferences in the selections of shirts and trousers (women's styles are less clearly marked this way and have not changed). In 1965 the stereotyped styles had been most consistently adopted by the residents of the Jat side of the village, except for the Brahmins and the Muslims. It had often seemed to me that the clothing and naming practices among Harijans and Sweepers showed a certain skepticism in regard to the way they would be treated if they claimed the full equality they were due in strict accordance to the religious principles alone, a sense of tension between reliance on the ideas of kinship and ideas of religion. In 1978 the tension was not gone, but the balance seemed to have shifted toward greater confidence in the religious framework. I did not offer these impressions in detail for discussion, but I did ask if more villagers had become more interested in Sikhism more of the time and to this everyone agreed.

Finally, I am sure that villagers would expect me to include the new school as a further indication of the increased importance of the Sikh religion and perhaps of religion in general. This is not because they believe the school to be a sectarian institution in a Western sense. It is not. But Panjabi is the language of instruction, respect for education itself is considered a basic value of the religion, and the aims and values of education largely overlap those of the religion. It is not just that Sikh *means* student; it is also that the religion does not advocate knowledge of some other world but of the fundamental basis of life in this one.

RESURGENT SEPARATISM: 1978–1983

The *Wall Street Journal* of December 17, 1982 carried a front-page story headlined: "India Throbs Again With Mass Protest, This Time by Sikhs." The article began:

> It is the same every afternoon. The speeches begin at 4:30 inside a marble assembly hall near the Golden Temple, the holiest shrine in the Sikh religion. For two hours, Sikh priests and politicians inventory their complaints to a crowd of men in orange turbans—orange for protest.
>
> At 6:30, fired by religion and political outrage, the men march to the central police station, and at 7 P.M., as planned, the police arrest them. By 7:30 it is over: The temple is sending over dinner for those jailed, the police are putting away their riot sticks, and another group of men . . . has begun arriving from the villages for tomorrow's demonstration.

This is all correct, but it necessarily distorts what is happening by detailing only the religious symbols in the protest and not the "complaints" that filled out their content. In fact, the underlying anatomy of the recent disturbances is precisely the same as that of the disturbances that led to the division of the state in 1967. The manifesto of the recent events, accepted by virtually all parties, is the "Anandpur Sahib Resolution" of the Akali Dal adopted October 17, 1973, during the Emergency. There was not much discussion of it during the Janata period, but it became a major focus again soon after Mrs. Gandhi's return to power.

The resolution is brief. It lists two principles, four aims, ten programs to secure these aims that are mainly religious in nature, and seven objects it intends to achieve that are more political. These latter suggest the main practical grievances. The first is to add to the state the Panjabi-speaking regions that had been left out in 1967 in spite of the clear acceptance of the linguistic rather than religious criterion. These regions include Chandigarh, Sindh and Kangra, and several sites quite important in Sikh history. The second was that in this new state as in other states central intervention should be strictly limited to defense, foreign affairs, post and telegraph, currency and railways. In effect, this would be a reversal of the present Indian concept of reserved powers: they would go to the states and not to the central government. This was expanded in another provision which called for the Indian Constitution to become "Federal in a real sense" and to assure that "all States are equally represented at the Centre." The other provisions mainly concern the elimination of discrimination against Sikhs and other minorities in government service and in civil life outside of Panjab.

The more concrete grievances have been explained outside the resolution in broader discussions of federalism. They have mainly revolved around four complex and interrelated topics: prices of agricultural inputs in relation to support prices, canals, industrial development of Panjab vis à vis other states, and employment.

The agricultural pricing issues that emerged after 1978 were almost exactly the same as those that had caused resentment before 1967. The relatively favorable balance between costs and support prices of 1978 was upset beginning in 1979 by the rapid escalation of world oil prices. Indian prices for fuel, fertilizer, equipment, and cement could not be insulated from world trends. As a result, announced input prices began to go up at rates of 20 to 30 percent a year. Shortages became endemic and black market prices went up much more. Rationing of such items as fuel and cement was occasionally imposed beginning in late 1978, but did little good. Yet support prices went up just 5 or 10 percent per year. On top of this, the very poor wheat harvest that winter of 1980/81 led to early projections that the central government, buying at the support price of about Rs. 127 a quintal, would not be able to meet its procurement goals. Prices to retailers went up to Rs. 200 per quintal in some cases. The central government's response was not to raise prices (not even to the Rs. 147 the state had originally asked for as the support price that would let farmers recover their costs) but rather to impose a ceiling on the amounts of wheat brokers could stockpile and at the same time to reimpose food zones even more severely than before. The *Tribune* of May 24, 1981 said that every district in the state was a wheat zone; farmers in surrounding districts could be arrested for taking wheat to Chandigarh—to say nothing of shipping to Delhi or Uttar Pradesh. Since government procurements went in the same network for city distribution as before, the farmers once again had no choice but to conclude that their interests were being sacrificed in order that the Congress party might maintain its constituency among the urban poor.

The canal issues were also basically the same as before. With the division of the state they were simply exposed more clearly. What had formerly been hidden within one state budget now became a matter of explicit arrangements between two states or more. The issues concerned both payment for canal construction within Panjab by states that would receive the water and the allocation of water itself. All these negotiations had to receive the approval of the central government, and it was generally felt that Panjab was not getting its due. One particularly contentious project was the Beas-Ravi Link Canal that took water to Haryana. Haryana was to receive 50 percent of the water under an agreement negotiated in 1978, but it was widely believed that the state would have been satisfied with 40 percent. The agitations of 1982 began when Mrs. Gandhi broke ground for the project.

Another major dispute concerned the control of the Bhakra headworks. The Bhakra had been built by Panjab after 1947 out of state development funds, yet after 1967 it was run by officials of the central government. This in itself was of no great consequence, but in the important spring planting seasons there were constant shortages in both water supplies and power generation from the dam and repeated instances of poor coordination of its power production with coal arrivals to a large steam-turbine power plant in Bhatinda, in northwest Panjab, which should have been able to smooth out the load in low-water periods. Since coal shipments were also a central monopoly, it was easy to see the overall power problem as the direct result of a pattern of central government inefficiency and indifference, which Panjabis quite naturally thought they could improve upon on their own.

The industries issue had begun to surface in 1978 and grew in part out of the migration patterns that had already become well established in Shahidpur as elsewhere—the inflow of laborers and the outflow of those seeking nonfarm salaried work. The trouble was that salaried work did not become available in proportion to the demand and there was from the local point of view no good reason for it. The central government controlled the growth of industries through licensing and Panjab had never been able to develop its industry in proportion to its agriculture or as rapidly as other states with weaker agricultural bases. It did not have spinning mills for its cotton (and many felt that its cotton prices were held down by lobbying from the Bombay mills that did do the spinning) and while it was a major sugar exporter, it did not have enough sugar mills to provide for its own local consumption. There was only one tractor factory in the state, founded in 1976, and even though it sent tractors to other regions, most Panjab tractors were made elsewhere. The widely publicized fact that the indexes of prices for agricultural commodities were increasing 5 to 10 percent per year less rapidly than for industrial goods only increased tensions. Panjabis widely believed that they were being economically exploited by a kind of internal mercantile system and that this was getting worse.

The employment issue overlapped that of industrial development, but it had an additional aspect that linked it to the issue of Sikh survival in a very interesting and dramatic way. Many Panjabis quite naturally connected the difficulties that well-trained Panjabis were having finding work in the state with well-known patterns of migration outward—to other states, to the Middle East, and to Europe and America. This in turn they naturally contrasted with the pattern of in-migration of agricultural laborers who were, of course, generally both non-Panjabi and non-Sikh. It was not hard to see these trends as a fundamental threat to the continuity of the Sikh cultural community (their phrase) in Panjab. Many had no difficulty in taking this one step further and seeing the pattern as the result of a deliberate design.

Added to these problems were three issues that were more clearly religious. The first concerned the Nirankari sect who affected to be Sikhs and claimed the Sikh scriptures as their own, but who overtly rejected the accepted historical exegesis and, from the Sikh point of view, insulted the gurus. In addition, they insisted on holding meetings in Amritsar, near the Golden Temple. On Baisakhi Day of 1978, a group of Sikhs marched on a Nirankari meeting in protest and the Nirankaris opened fire and killed some thirty-six of them. The main suspects in the case were eventually acquitted because of contaminated police evidence. The head of the Nirankari movement was cleared of complicity without being indicted and continued to preach in the state, backed by strong statements of central government figures to the effect that they expected the state government to protect religious freedom. Eventually, in April 1980, he was assassinated. In contrast to the seemingly lax handling of the murders of the Sikhs, the investigation of the assassination was vigorously undertaken by the Central Bureau of Investigation, and the papers were full of stories of torture of suspects and the like. Repercussions continued into 1983.

The second major issue concerns the demand for an all-India gurudwaras act. Since the SGPC is presently established by a Panjab act, its scope is limited to that area. A national act would extend SGPC purview to all the gurudwaras in India, assure that funds from gurudwaras in other states were channeled into the single gurudwara system, and allow Sikhs all over India to participate in the SGPC elections. Without such an act, Sikhs who left the state were in a sense lost to the main organization of the community. This was directly connected to the threat seen in the outward movements of Sikhs to other areas, especially India's major cities.

The third issue concerns Radio Golden Temple. As a logical extension of the practice of broadcasting the Sikh morning service and evening talks over village public address systems, an Akali group calling itself the Radical party announced in November of 1979 that they would begin broadcasting the ceremonies from the Golden Temple each day by radio. Radio licenses are however a central monopoly, and the government would not grant one. Broadcasts started anyway on November 15th, with a radio said to have 25 watts and a range of 200 to 250 meters. This led to a wonderfully droll comedy in an indefinite number of acts that perhaps only a Panjabi could appreciate fully. As the author of the project said, "It was symbolic but real." The persistence of the central government in refusing the license and trying to stop the broadcasts was taken as clear evidence that its action was motivated by pure prejudice and not by any rational considerations at all.

As of 1983, the package of grievances brought together under the heading of Sikh religious ideas contains at least one logical extension of every point that was in the package of 1965, with appropriate policy recommendations to match. The central government has generally chosen to represent the movement publicly as communal or sectarian. A more constructive approach would be, as before, to understand first that these religious ideas are deeply important to those who accept them, and then to realize that because of this importance they would not be used to state trivialities. Indeed the Sikh religious ideas are used to build and reflect consensus on a number of quite concrete problems that are of the most immediate practical importance, and the only response that will be effective in the long run is to turn to these problems and solve them.

CONCLUSION

Like the kinship ideas, the basic ideas of the religion remained constant throughout the period of the Green Revolution, even though patterns of consensus on the relative importance of some of the principles and on their applications were in continuous change. Yet just as with the kinship ideas, it would be a serious mistake to say that since the religious ideas did not change they were not important, or to say that because they framed more practical and not exclusively religious issues they were not also valued in their own right. The religious ideas were important in almost all the ways that systems of very low information potential can be important, ways that complemented but did not duplicate the importance of systems of ideas which more directly governed ecological, economic, and managerial activity.

No set of ideas as abstract, tightly organized, and simple as the ideas of the Sikh religion can direct people or groups in the conduct of such complex changes as occurred in Panjab between 1965 and 1978—or in the years before or since. The ideas had to be fitted piece by piece to the situation—that is, to the systems with higher information potential. If they had not been coherent, easy to grasp, and already widely held, they could not have been so used, but the uses were not by any means foreordained.

It is precisely because the religious ideas had to be actively applied to the situation that their role in the Green Revolution in Panjab cannot be separated from the development of the Sikh religious organization, which in turn cannot be separated from the history of the region. Partition created the basis for the new state. The religious organization argued for it and formulated what became its new policies and goals. The policies reflected genuine economic needs but, equally important, they unquestionably also reflected widespread personal acceptance of the ethical and intellectual power of the religious ideas as such, and the desire to preserve and defend them. The same religious organization that created the new government later formed a vigorous and independent platform for criticism of government actions which fell short of those same religious ideals, and it has continued to do so. The uses change; the ideas and the manner of use remain the same.

CHAPTER 8

Parties

The third and simplest village system of generally accepted cultural ideas of low information potential is that of "parties." Like the ideas of religion, this system did not change in its fundamentals between 1965 and 1978. But it did change in its applications and these changing applications were important. Throughout the period, the concepts of parties provided the framework used to mobilize support for and against public policies, to group opposed individual interests into organized pressure groups, and to assess the relative strength of such groups and the power they wielded with extraordinary exactness.

Where the ideas of kinship and of religion both define systems of cooperation, utopian ideals of society if carried to possible logical extremes, the party ideas define a system of conflict that is considered to reflect the more corrupt or selfish side of human nature; where the kinship and religious ideas both define relationships that are explicitly commended, the party ideas define relations that are explicitly condemned; where the religious and kinship systems both define ideas that are represented overtly in public manners and ceremonies, the party ideas define relations that are said to give such events their hidden or secret meanings. Yet this set of ideas is itself as completely well known and public as any of the others. As a basis of political activity, it is fundamentally unlike either American or European ideas of political parties in its content and logic, but it is by no means any less coherent or useful.

THE BASIC CONCEPT

The Panjabi term "party" (*pāRTī*) comes from the English, but is now thoroughly naturalized. By January 1965, I had collected in passing most of the bits of information on parties that I have presented so far. But whenever I broached the topic for systematic discussion, people stopped talking or denied that parties existed. Parties were clearly not things you told outsiders about, and by the very fact of asking I assigned myself the outsider's role. Still, what information I had did not seem to form a single plausible model, so I waited for a more opportune time to ask for an explanation. Such an occasion finally presented itself just before the SGPC election. I had been present one morning in a farmer's house during some discussions of party matters and as usual kept quiet. Naranjan had been there as well, and afterwards invited me to lunch. After eating, I asked Naranjan to explain what parties were. He immediately said there weren't any in the village (choosing to forget the morning discussion). I explained very carefully that I did not want to know *who* they were, but rather to understand *what* they were, since what we called parties in the United States seemed to be different. We did not consider that all the people in one party were friends and that those in the opposite party were all enemies. The ideas of parties and friendship were quite different for us and unrelated. Similarly, we did not associate parties with court cases, or attempts to get land, or indeed with personal conflicts of any sort. Nor did we consider that a person was bound to agree with all the positions taken by his party and disagree with all the positions taken by the other. Our system of local parties, formally designated, built up into a national hierarchy, but Naranjan and others had said that in Panjab there were no local parties or local party leaders. Finally, we did not generally think of the winning party in an election as "dominating" the losing party in the manner of a military victory, or that the losing party should therefore keep quiet. Rather, we considered that it was still proper for the minority to voice its views and for everyone else to listen to them.

Given all this, Naranjan agreed to explain and said that the parties were enemies because they disagreed on some issue which was a matter of principle. I could see why a disagreement over principle might produce enemies, not just disagreement, and it might explain why they struggled for dominance, but it did not explain why those in each party were all friends to one another, why parties were secret, why they fought court cases, why they dominated panchayat activities, or why there were no local parties or local leaders. I asked if he could draw a diagram. He said yes and took the pencil and paper that I had folded to a blank sheet. In the middle of the empty space, he very deliberately drew a single vertical line, nothing more. The diagram made even less sense to me than the original remark about issues and principles. I had been expecting some variant of a pair of descending branching hierarchies, with the whole structure representing the national parties and steps representing state and regional units, down perhaps to the village level groups. Then issues or principles could be the platforms of groups at one level, adopted in some way. Had my request been understood?

I was afraid that if I drew such a figure myself and Naranjan agreed, it would not be possible to know if he was agreeing simply out of politeness. So I first drew about a half-dozen horizontal lines, one above the other like a layer cake and asked if that might be parties. No, Naranjan assured me without hesitation, that was not parties, that was *jātīs*. Confident an incorrect prompt would thus be refused, I then drew the hierarchy and again asked. Again, I was assured without the least hesitation, that was not parties. His drawing was parties. I tried to explain that it didn't make sense: I couldn't connect it to the idea of enemies, or domination, or elections. Naranjan took back the page and added the rest of the elements that appear in Figure 8.1.

There is still no hierarchy. When I explained my own alternative diagram, Naranjan said he understood it but assured me it did not apply. There were no local parties. There were only national parties which everywhere opposed one another on the issue. The line represented that opposition, absolute and unqualified. Seeing that the issue was "yes" as against "no" did seem to make sense. However, many issues were being discussed in the election just coming up. When I explained this to Naranjan and gave some examples, he said that was not correct; the issue was just one: who would win the election. Each party says "yes" to its candidate and "no" to the other and only one candidate can win. The logic was stunning and for a moment I hesitated. Still, the picture wasn't quite ironclad. I objected: a man is not *really* an issue or a principle. Yes he was, Naranjan explained, because each man took positions on a series of issues and related them to one another and these relations as well as the oppositions between the opponents were (as I have already noted) explained in terms of general principles. You could not support the man without supporting his positions and principles; you could not support those principles without supporting the man. Thus in the last analysis, it was yes or no on the principles and yes or no on the man. Just yes or no, win or lose, no middle ground and no shading.

This sense of an "issue" did indeed explain the dramatic all-or-nothing op-

Figure 8.1
Parties

Party:	A	B
Issue:	Yes	No

position between the groups, resulting in dominance, but how did it account for the lack of hierarchy? The basic idea was that since the parties were defined by opposition, anyone who supported the major leader was in his party and the only thing that those who were in his party needed to express was support. No one held at the outset that the two leaders disagreed over absolutely everything, but what was agreed on did not require debate. The major leaders simply ignored their areas of agreement. They were not at issue. To set up an issue was to set up opposition.

Logically, an important local person active in politics would have only two options for overt action, and neither would allow him to set himself up symbolically as a local leader in a Western sense. He could endorse one of the national leaders and reject the other(s), in which case he would be "of" the party of the person he endorsed. Or he could announce a difference of opinion with all the national leaders, in which case he would still not be a local leader but rather would be attempting to set up yet another national party. One cannot both endorse someone and disagree with them, no matter how trivial the disagreement might seem to a Westerner. In effect, the idea of a local leader is a contradiction in terms. Leaders were national by definition because the principles they were speaking for were universal. There were but two leaders because any third party, opposed to both major parties at once, would automatically be swamped in a contest. This was also why I was told both that there were no local parties and that the local patterns of leadership that did exist in fact had to be "secret," that is, without publicly acknowledged recognition. With this and what I already knew, the whole picture fell into place.

The next questions, had I needed to ask them, would have been "But why are the opposed parties present in each village?" (rather than, perhaps, one party in one village and one party in another) and "Why are these parties connected to fighting for land and other goods?" By then, however, the answer was clear. It is not that the national parties somehow generate disputes within the village, but rather that disputes within the village are viewed as inherent, arising both from the necessity of competition and the unavoidable facts of "love" and character. There is never quite enough to give all people what they need. At the same time, certain people will have a natural affection for each other and agree on issues, while others will just as naturally dislike each other and disagree. There is no accounting for such impulses, in the local view, but they are held to be inevitably mutual. Given this, it is logical that each local person will seek help outside the village to support his own case, and it is at this point that he becomes willing to trade his own support to outside leaders for it. Parties grow outward, from the personal likes and dislikes and from the clashes of interests of the villagers. They are not ideological and do not grow downward from "power brokers" or any such type. The local supporters of each national leader are supporting him because it is in their interest to do so—because he is articulating a case suitable to their positions and will directly or indirectly aid their personal causes. The national parties, conversely, are not ideological groups but constellations of support put together by specific individuals.

One of the sources of confusion among Western observers of Indian politics is the failure to realize that the regional and national groups that go by the name of political parties, such as the Congress and the Akali Dal, are not actually parties in this sense. Rather, parties operate within them just as they operate within villages. This is the reason one often sees two or three Congress slates or two or three Akali Dal slates in any given contest—generally defined by a separate leader, such as the Sant Akali Dal or the Indira Congress. Within the village, at any point in time, there will be two parties, but only the dominant party will be able to make its views known publicly (although covertly). It will present these views as the views of the entire village. Supporters of the other group will in most cases either keep their opposition to themselves or, if the difference is at the point of breaking into the open, withhold endorsement by boycotting the event in which the views are expressed, be it a small local ritual, a large regional political rally, or a wedding ritual. Similarly, in the Akali Dal or the Congress, the dominant group will purport to speak for the whole on public occasions, but everyone will know some members of the leading body who in fact support other views. Depending on the circumstances, these minority members may go along in silence, absent themselves from the proceedings, or they might even contest the activity in question. This last is a most serious move, functioning much like a vote of no-confidence since it logically rejects the accepted understanding of who is dominant.

These patterns, all clearly following from the idea of parties, make up what Westerners have often described (and decried) as the "personalism" of Indian politics—the constant shifting of alliances which produces coalitions that on ideological grounds are at least unlikely and often bizarre. They are also what lets precisely the same village group appear as supporters of Sant Fateh Singh as opposed to Master Tara Singh in an SGPC election, but as supporters of the Congress party as opposed to the Akali Dal in an election for state office, even if Sant Fateh Singh is a major figure in the Dal and opposed to the Congress. What counts is which of the two contenders for office best represents the local interests. It is also why Indian public media, from village gossip to national newspapers, consistently report events of political significance by describing not only what is said by the major speakers but also who else attended, what they might have said, (although the fact of attendance alone usually speaks for itself) and occasionally who is not present. When the positions presented and the array of those who support them (with the chain of further support that they entail in turn) are taken together, the long-term significance of the event can be quite sharply and subtly set out.

The political significance of Sikh religious gatherings is derived from the party ideas in precisely the same way as is the significance of overtly political events. The religious principles provide the overt bases of the claims for moral support; the party ideas provide the covert significance of the pattern of attendance and the nonreligious issues alluded to as illustrations. Thus the full political power of the religious ideas comes not from the religious ideas alone, but from their use in combination with the party ideas.

Within the village, the idea of party is seen to underlie, or potentially underlie, all public acts. Villagers can give descriptions of the village to this effect. If, for example, one meets a relative on the street and uses a kinship term for address in the proper manner, all will be considered well. But if for any reason one neglects the proper greeting or perhaps doesn't deliver it in quite the usual outgoing way, it is considered to be possibly significant of hidden party opposition. If one attends a wedding of a relative, all is well. If one does not, it is considered to signify support of the other party. If one attends a reading of the Guru Granth Sahib sponsored by a villager, all is as it should be; if one does not, again it may signify party opposition. This is a major reason why villagers, even those who have no interest in parties themselves, are very careful to follow the formal rules of decorum connected with the kinship and religious ideas. It is also why the only defense against such entanglements is the claim of absolute ignorance of what parties are. Miriam Sharma's recent account (1978) of party activities in a village in Uttar Pradesh gives an excellent sense of the way village affairs can look when seen through such party ideas consistently applied as the main conceptual framework, although it neglects the point that not all people apply them all the time and that other ideas can be given priority (and the party ideas set entirely aside) with equal ease.

Strictly speaking, to someone who knows what the symbol means, the idea of parties was accurately set down in Naranjan's original diagram, the bare line representing the fact of opposition itself. It is difficult to imagine any system of ideas simpler than that. At the most, the irreducible minimum set of separate ideas would be the ideas of opposition and of an issue represented by Naranjan's second diagram.

PARTY DYNAMICS: 1965–1978

In 1978 the dominant group in the village supported the Congress party, even though Panjab had been one of the states where Indira Gandhi's Congress had been resoundingly turned out of office in 1977. This did not reflect any misunderstanding of Congress's strength. Villagers said that at the time the Akali party was first in the state, with Ajanta (the local version of Janata containing Akali, Jan Sangh, and some former Congress groups) second, and the "Marxists" (or "Left Communists") third. Congress was a weak fourth, with about 15 of 114 seats in the Legislative Assembly. In the 1977 election, the other three parties had formed a "United Front" against it and in 1978 the Front was still strong, although even then it was beginning to erode in the face of Janata inability to curtail increasing inflation and avoid some impending shortages in important goods and materials, such as diesel fuel.

The obvious question, in view of the idea that one supported national parties to gain help for local needs, was "Why support such a weak group?" The answer, as one should expect, was that the reasons were local. In fact, I was told, some of them concerned the building of the school. The story of the school illustrates both the continued viability and importance of the party ideas and the way they are intertwined with relationships formed on the other available bases.

The School: Parties in the Region

I was told the history of the school on two major occasions, with more information filled in during numerous other conversations. The first occasion was shortly after I had returned and the story was phrased in terms of the formal public events that led up to building it and the sources of support. The second main account was given to me in a more private setting about six weeks later, when many more understandings had been reestablished. This account dealt with parties.

The second discussion of the school began in response to certain of my questions about the headmaster of the nearby private school, Jasminder Singh. Jasminder had been supported by the dominant village party in 1965, but had been allied with the Communists (Right Communists, I had thought) and not the Congress. The Communists had been a major group in the area in 1965 and had grown still more in strength during Indira Gandhi's period of Emergency when they and the Indira party of the Congress formed the majority coalition in the state. But they had been virtually wiped out in the 1977 elections. I was curious about the mechanism by which a party could disappear in this way and about the effect it would have on the individuals involved, and it was really just to explore this that I brought up the headmaster. I was first told that the headmaster was not Communist and had never been. He was Akali, and had been then as well. He was, I was told, "always Akali." Since I knew that the dominant group in the village had also been Akali in 1965, but was not now, I naturally asked if he was still supported by the village group. The answer was that he was not. Indeed, the hostility between the headmaster and the local group was a major reason that the village group was now Congress. When I asked how this had happened, the school was part of the explanation.

There are some inconsistencies and gaps in the stories, but this is to be expected of what are in effect two distinct series of events that were seen in different contexts and visible to at least partly different sets of persons. For the sake of coherence it will be best to begin the story in 1972, when the new panchayat was elected with Indar Singh as sarpanch.

Indar Singh's main opponents were two men who had been in the majority party on the previous panchayat, which Indar Singh himself had supported. Siani Singh was Jat, Jagir Singh was Harijan. Jagir Singh had been sarpanch during my first study and was running for reelection. Siani Singh was not running for reelection himself but was supporting Jagir Singh. Neither man had much formal education, had ever lived anywhere but in Panjab (mainly in Shahidpur), or spoke any language but Panjabi. Both, however, had well-established regional reputations for their political abilities. Siani Singh, in fact, had been the person most important for me to meet when I first arranged to come to Shahidpur in 1964. He was respectfully described by Rajindar at the time as a "very resourceful man." He had very little land and spent much time out of the village at local courts and offices inventively and seemingly tirelessly pursuing the interests of various villagers and other friends. Yet he seemed so thoroughly nonideological it never occurred to me to ask what his formal political affiliations might be. Jagir Singh farmed a moderate amount of land held on tenancy and mortgage and was a regular participant in political meetings and religious gatherings of the Left Communists in the area, where he often lectured and sang in a duet with the young son of the village *granthī*.

Indar Singh was quite different. In 1965 he had been moderately interested in politics, but only among many other things. He had a fair amount of land and concentrated on farming it. No one mentioned a regional reputation for "resourcefulness" like that of Siani Singh, but he was apparently very well regarded and respected in the village. He had enlisted in the British Army while he was in the tenth grade in school and had attained the rank of sergeant major before his retirement. He spoke and read some English in addition to Hindostani, Urdu, and Panjabi. His interest in and regard for education was apparent from our first acquaintance. He insisted on it for his sons and was consistently among the first to adopt innovations he had read about.

The best way I can convey a sense of Indar Singh's character is to recall that when he was harvesting his crops, especially when fresh sugar was being poured out of the boiling pans to cool, poorer villagers would regularly join the group and lay a cloth out on the ground. Into this, some of the crop would be put—with nothing direct being said. About the third time I saw this was one winter evening as the family was making raw sugar. After the visitor left I asked if he worked for the family. The answer was no, he was a poor person of the village and there was an obligation to help such people. I asked if all farmers gave such help. Netar replied, in English: "No, there are hard families and soft families. We are a soft family." As with Siani Singh, there was no occasion to ask Indar Singh's formal political preferences in 1965 and I have no idea that he had any, apart from the fact that he supported Sant Fateh Singh in the election of 1965.

It was still painfully clear in 1978 that the 1972 election had been bitterly contested. The vote was close and there had been irregularities. As a result of his defeat, Jagir Singh left the village. Siani Singh has apparently not spoken to Indar Singh or his close supporters since.

Apparently Indar Singh had a school in mind when he was elected. In 1973 the village was host to a series of tournaments and games, a kind of folk-olympics. Prizes were given out by Mrs. Sharla Prashar, state Welfare Minister (and later important in the Congress party, Indira group). Also present was Sant Ram, whose family had moved from Shahidpur to Sirhind and founded a substantial business.

Sant Ram put up a prize of Rs. 200 for the *bhangRā* (a Panjabi folk dance) contest and on this occasion told Indar Singh that if a school was built, he would donate a large room. This was apparently the first commitment.

To obtain a school, it was necessary to establish interest and need and to forward the request through a Member of the Legislative Assembly. The local MLA was of the Akali Dal, so the villagers took their letter of request to Jasminder Singh, who was prominent in local politics and had been elected to the panchayat samiti in 1965. His endorsement of the request would have greatly simplified the villagers' problem. As it was told to me, when he was handed the letter of request as he sat at his desk in his office, he simply folded it up and threw it aside. He would not support the villagers' request and advised them that they could be sure that the local MLA would not either. Because of pressure of time, I could not find Jasminder to hear his version of the story, but I had known him fairly well previously and found this entirely in character. He had worked aggressively to build his school literally from nothing and was quite ruthless in extracting support for it whenever he had any leverage by which to do so. A government school, with free tuition, would be bound to strike him as a threat. He also believed more literally and consistently than most that party considerations were pervasive in all areas of life and were fundamentally self-serving.

The panchayat next took its case directly to the Education Minister of the state, who was Congress, with the help of, as Netar and Indar Singh described it, "some other" MLA, with a motion of the hand toward the north. The entire panchayat met with the Education Minister in his office in Chandigarh in March 1975 and made its presentation. In reply, the Minister said he would come to Shahidpur on the second of April 1975. If, at that time, he saw that a school was necessary, one would be built.

On the appointed day, as I was told, there was a great *bandobast* ("gathering") in the village, with a stage set up near the village gate, a public address system, tenting and space for a large number of visitors. On the stage, chairs were provided for the important guests. Photographs were taken to record their presence. Among them were Ravi Indar Singh, Speaker of the Vidhan Sabha (upper house) of the Panjab legislature (Ajanta party); Raja Singh, MLA (Congress, from Morinda); Kanshi Ram, President of the Congress party central committee of Ropar district; and the headmaster of a private higher secondary school in Chamkaur. The latter had run against Jasminder Singh in several contests around 1965 and had always been associated with the Congress. I suspect he had something to do with the support of the "other" MLA. Jasminder Singh himself was conspicuous by his absence. Nongovernmental visitors especially mentioned included several relatives of the aforesaid Sant Ram. One of them was his *patija*, Professor Jot Ram, who then lived in Sirhind but who had previously worked as an engineer in Rajasthan and who had been a classmate of Indar Singh in the primary school of Shahidpur from 1918 to 1922. Anant Ram, Jot Ram's father, also came. Anant Ram, Jot Ram, and Sant Ram not only reaffirmed Sant Ram's initial pledge but substantially increased it. The school would be built in the names of Bihari Lal, the father of Anant Ram, and Sant Ram and his sister, Jai Kaur. Bihari Lal had been born in Shahidpur about 1865, Jai Kaur about 1875.

Villagers said that about five thousand people came from surrounding villages to attend the affair. In the party context, this was an impressive show of support.

When I was first told about this meeting, all that was said was that when the Education Minister spoke, he said the school was necessary and when Ravi Indar Singh spoke, he said he would provide a government grant of Rs. 10,000 for the building. Later, from remarks in passing about other matters and from what was told to me by the secretary of Ravi Indar Singh when I visited his office in Chandigarh, it appeared that it had not been by any means just a matter of announcing decisions already taken. Offstage, there had been a genuine review and discussion of the situation with those involved, and the final decision was made on the spot. The officials had to assure themselves of local interest and need. The size of the crowd attested to both, as did conversations with representatives of surrounding villages. At the same time, promises were made firm about the support in the names of Bihari Lal and Jai Kaur and the panchayat promised to set aside the land of the former cremation ground and to raise the additional money. In short, the *bandobast* provided the public occasion for a whole group of interrelated underlying agreements that established the school and determined its size and character.

In addition to the Rs. 10,000 from the Speaker's "Dictionary Fund," major donations included Rs. 6,000 from Sant Ram, a total of Rs. 8,843 from Jot Ram, and Rs. 5,000 from Giani Gurmail Singh (the same who had contributed to the village gurudwara). The panchayat assessed the village on the basis of Rs. 1 per *bīghā* of land, Rs. 5 per turban, and Rs. 10 per house. "Some gave and some did not," as noted, for a total of about Rs. 3,900—over and above the three acres (1.2 hectares) of cremation ground. The salaries of the one headmaster and six teachers were to be paid by the government.

Nine months later the first classes met, using the village primary school building. Construction on the new building began in 1976 and classes began in the new building in July 1977.

When I was first told about the 1975 *bandobast* and shown the photographs, several people stressed that large numbers of policemen were present. It was said as though it signified something, but what this was only became clear in the later discussion of party ideas. The headmaster apparently had not stopped at withholding support but had actively worked to prevent the meeting by saying that there would surely be violence, in fact "a firing" (meaning the use of firearms), so the officials should not go. The presence of the police was taken as proof that he had circulated such a threat and as an indication of the political courage and depth of support of the officials who ignored it. Of course, there was in fact no violence of any kind.

In April 1978, when the school building was completed, a second *bandobast* was held to open it. Most of the same people who had attended the original event returned along with some additional prominent regional figures. These included several members of municipal committees of nearby towns and the president of the Ajanta party for Ropar district. Jasminder Singh was still absent. The headmaster of the school at Chamkaur was again present.

While Jasminder Singh's opposition "and similar other reasons" were cited in 1978 to explain why the dominant village group supported the Congress party and not the Akalis, I don't believe there was any intent to suggest that they were supposed to explain the internal village conflicts as such. These, as usual, grew from below and involved conflicts over land, as they had in 1965.

Land and Labor: Parties within the Village

In 1965 a comparison of land and population in the two factions showed that by and large they were composed of families with opposite and opposed needs, on two extremes of what appeared to be the more balanced situation represented by the village averages and by those not in parties (Leaf 1972:221–223).

In 1965, 21 families had been involved in the dominant party and 25 in the weaker, with a total of 421 people of which 373 were residents. In 1978, from remarks in passing, direct description, and observation of behavior, I was able to compose a list of 11 families supporting each party. Their total population was 227, of which 190 were residents. Almost certainly there were others equally active that I missed. The likelihood of this is enhanced by the fact that my list includes no landless families in the minority party. Nevertheless, bearing in mind the risks, it is better to set down what is known than to say nothing.

The dominant party in 1965 involved a core of landowners with comparatively small amounts of land owned per capita and relatively large amounts of land held on more insecure arrangements, shown by a high ratio of mortgaged land to owned land. These farmers were then allied with a relatively large number of absolutely landless villagers. The party's affiliation with the external Communists, favoring lower landholding limits and less compensation, was consistent with the interests of both groups.

The minority party centered on a group of landowners with almost twice as much land per capita and much more land held on outright ownership. Where the landowners of the dominant party held 39 percent as much land on mortgage as on ownership, this group held only 9 percent. In fact, this group had so much land that it didn't use all of it for income; on the other hand the income of those allied with the dominant party exceeded the income obtainable from land they owned (97 percent compared with 113 percent, respectively). Finally, this group of owners was allied with a slightly smaller group of nonlanded families, evidently those to whom the farmers gave direct employment, and a few long-term partners and tenants. Twenty-six percent of the supporters of the minority party (47 individuals) were from families without land, compared to 28 percent (64 individuals) for the dominant party. The minority group's support of the Congress party, which at the time was considered to be more cautious in its advocacy of land reform and more protective of established interests, was considered to reflect the group's interests in maintaining secure title to its land.

Table 8.1 gives the figures for 1978 and includes the main index ratios from 1965 for comparison. Despite the shifts of membership back and forth,

Table 8.1

Landholding and Population of Shahidpur Party Groups, 1978

Statistic	(A) Dominant party absolute ratio	(A) Dominant party decimal ratio	Total village		(B) Minority party absolute ratio	(B) Minority party decimal ratio	Decimal ratio[a] 1978	Decimal ratio[a] 1965
Land held on mortgage / Total land owned	0.6 / 156.45	0.00	37.32 / 1832.46	= .02	.01 / 451.13	0.00	0.00	4.33
Total income / Total land owned	206.52 / 156.45	1.32	1907.23 / 1907.29	= 1.0	461.13 / 470.09	.98	1.35	1.17
Total land owned / Owners and households	156.45 / 61	2.56	1907.29 / 366	= 5.21	470.09 / 97	4.84	0.53	0.59
Total income / Owners and households	206.52 / 61	3.38	1907.9 / 366	= 5.21	461.13 / 97	4.75	0.70	0.68
Total income / Total party population	206.52 / 93	2 22	1907.79 / 977	= 1.95	461.13 / 97	4.75	0.47	0.67

Note: Dominant party includes 11 families and 93 residents; total village includes 68 families and 366 residents or owners; minority party includes 11 families and 97 residents.

[a]Decimal ratio is computed by dividing Statistic A by Statistic B.

the constancies are striking, although there are also some obvious and predictable changes. The dominant party still had more income than could be obtained from the land it owned (132 percent compared to 113 percent in 1965). The minority party obtained income equal to slightly less than what its owned land would provide (98 percent compared with 97 percent in 1965). The dominant party had substantially less owned land per capita than the minority party. It held 2.56 *bīghās* per capita in 1978, compared with 2.66 *bīghās* in 1965; the minority party in 1978 held 4.84 *bīghās* per capita compared with 4.51 in 1965. Thus the amount of land per capita among landowners in the dominant party was 53 percent what it was among the minority party, compared with 59 percent in 1965. Because of the different rates at which the owners extract income from their own land, the gap between the amounts of income per capita is less than the land owned per capita—70 percent in 1978, compared with 68 percent in 1965.

The first and most predictable of the differences concerns the ratio of land held on mortgage to land owned. In 1978 this no longer yielded a significant difference because neither party had any land on mortgage at all—a reflection of the more general change toward landowning alone as a basis of farming. The second difference is that the spread of the different rates of utilization of owned land for income has increased slightly. The third difference is that the gap between the proportions of people from landless families, and hence between the overall levels of income per capita in each group, has widened. In 1965 the income per capita of the dominant group was 67 percent that of the minority group; in 1978 it was 47 percent. However, since this difference would probably be reduced by better representation of all types of families involved, it is best to take this as meaning only that the direction of the differences is the same and that the magnitude may have changed. More information is needed.

My analysis to delineate the interests underlying the village party alliances in 1965 related the land held by each family to the total family size. For 1978, since the returns for total family size were not really comparable, the calculations could not be made on the same basis. Therefore the data for both periods were recalculated relating land only to the family members in residence. This measure showed the same pattern of contrasting imbalance. In addition, however, it showed that for both groups in 1965 and for the dominant party in 1978, the party groups were consistently below the averages for the village as a whole. That is, families active in party matters tend to have less land per resident than do families that are not active. There are two interrelated reasons for this. First, those active in party matters (except the dominant group in 1978) tend to have more of their family members in residence. Second, they evidently rely more on agriculture alone for their income than is usual. This is what pressed them to compete and, given the very different balances between their human and landed resources, led them to organize this competition by supporting opposed economic and social policy positions in order to obtain the support they needed both within the village and without. Remember that in 1978 it was the Congress party, which the dominant party supported, that favored the more drastic land reform, while the Akali party that the minority group supported had become part of the "go slow" group.

In sum, there has been no discernible change in the content of the party ideas or in the emphasis placed on one or another aspect of them (for example, on the idea of dominance as opposed to the concept of principle), and there appears to be no major change in the context within which the ideas are chosen and used that can be attributed to the party ideas as such. The changes are due rather to new external circumstances: the opportunity for the school, the ebb and flow of family fortunes, and the changing pattern of external policies and programs through which to pursue family interests.

CONCLUSION

In my 1972 account, I described the kinship system, religious ideas, and party ideas as "conscious sociological models" to emphasize the points that they had in common and that distinguished them from the ideas of the ecology, division of labor, and the economy. Together they formed a kind of system, but they were not a "structure" in the sense of a single coherent framework with each element rigidly tied to every other. In fact, the system they formed was based squarely on the idea that they could not all be used "at once," to define a single recognized social situation. Rather, they were just as Fred Bailey described the concepts of tribe, caste, and nation in an area in Orissa—three "incompatible" structures that exist "in a single social field" (1960:7–9). Within a single structure one might have clearly defined "conflict," a definite relationship of opposition (or its opposite, cooperation), and the relationship between one person and another in one situation could be reliably inferred from their relationship in another. However between such structures nothing could be inferred in a regular way. In Shahidpur, if two people are in the same family, they are necessarily in the same *jātī*. If they are in two separate *pattīs*, they cannot be in the same family. If the one is in a higher-ranking *jātī*, he is necessarily (within the context of kinship) also in a higher ranking *pattī* and family—and so forth. Anyone knowing the ideas can make dozens of correct relational inferences from any one statement. But nothing whatever can be inferred from any kinship relationship to any party relationship. If two people are of different *jātīs*, they may or may not be in the same party; if they are brothers, they may or may not be in the same party. By the same token, nothing can be inferred between religion and kinship, or religion and party. They are simply separate matters.

Yet people know when and how to read an underlying party meaning into kinship usages, or into religious usages, and when not to, and they know when they can logically combine ideas from kinship and religion (as in speaking of the guru's patrimony) and when they cannot (as in avoiding mention of castes). If such relationships are not themselves part of a larger structure, where do they come from? They come from the way the separate rules of use implicit in each idea (what Wittgenstein [1965] would have called their "grammar" and what I have been calling their "logic") bear upon the actions of the individual, just in the way the shapes of a screwdriver and of a pair of pliers make it impossible for a person to use both at once in the same hand even while suiting them to complementary functions. Kinship and religion are both public, but in different ways, so they clash. Party is hidden, so it can be hidden behind either.

Kinship is, in theory, public because it is based on birth and marriage which are both public events, connected to occupations, and reflected in patterns of affection, cooperation, and manners (such as address and deference). Religion is public because it is, in theory, based on the voluntary association of a person with a teacher, which is demonstrated by attendance upon that teacher in his place (gurudwara) and by adopting behaviors that he recommends and gives significance to. Just between these two systems, there are certain rules that define certain choices that a user must make in regard to the ideas of the other system. The kinship ideas, in a kinship context, define property occupationally and entail recognition of inherited occupational differences. The ideas of religion use the kinship idea of a "patrimony," but define it as a set of ideas rather than material capital; they tie it directly to the idea of sustenance but not occupation. With the idea of a spiritual patrimony, the kinship idea of a brother can be applied to everyone equally—all people everywhere or at least all followers of the guru. With the kinship idea of property, however, there is a distinction between "true brothers" and brothers in what appears to be a more extended or attenuated sense. This means that there are occasions where kinship and religious ideas can be used together (such as within a family, under most circumstances), but there are other occasions where there is a definite clash of ideas, and the user has to select one or the other. One can define an occasion with the idea of *jātī*, or with the idea of the relations among all Sikhs (for example, by selecting the appropriate greeting and following through with the associated ideas), but not both.

The relationship between the two sets of ideas defined as appropriate for overt public use and the idea of parties that has to be presented secretly is the same, fundamentally, except that since parties are secret, one can use the idea of a religious relation and the idea of party affiliation, or of a kinship relation and a party affiliation simultaneously, rather than alternatively. In fact, it is hard not to use them this way. This is why if one uses any kinship term, it can and probably will be seen as also expressing party support; if one does not use a term that could be used (whatever it might be), that can be construed as expressing party hostility. Conversely, since parties are secret and therefore by definition have no formal or public mode of expression on their own, it follows that if one wants to express one of the party ideas, such as friendship or support, there is no other way to do it except by means of an idea defined in one of the "overt" systems.

There is no rule that says ideas can only be applied to a situation at just one level or in just one way, and indeed they generally are not. Each system is applied not like a conceptual coat of paint but many times at many levels to define what is on the surface and what is beneath it, what surrounds the action being engaged in and what makes up its elements. The systems are applied by individuals to objects around them to make up communicative situations, but such applications are inevitably recursive and reflexive, applying back to the user himself and in turn again to those who accept the user's attributions, again and again and again—off into a conceptual infinity like a reflection between opposed mirrors. The pervasive sense of reality thus conveyed is itself what makes the ideas important, not usefulness or convenience in the shallow sense in which an ideological slogan might be useful or convincing at a point in time. It is, by the same token, also what makes threats to any of these ideas so profoundly unsettling, for in the end they are not threats to a set of labels but to the very structure of reality; not to this or that interest or project but to the basic notion of regular relations on which interests and the promise of projects are based.

There has been no threat to the party ideas analogous to the threats to the religious and the kinship ideas, undoubtedly because the party ideas used in the village are substantially universal in South Asia and are indeed the language with which the other two threats have been organized. The efforts to weaken the kinship conceptions and the perceived threats to the religious ideas have actually reinforced the consensus on the party ideas. But in the long run the party ideas will never obliterate the religious or kinship ideas. Each kind of idea has its place; the villagers could not operate in a monolithic universe and undoubtedly would especially dislike a monolith made up of unending and unmitigated conflict.

The stability of each of these major models over the period of the Green Revolution has been achieved in a different way in each case, but in each case it represents a continuity of a moral framework that has not only been emotionally important to those involved but has been positively integral to the process. At different levels and in different ways, each of these ideas has played a role in organizing and mobilizing the resources that have created the more material results.

CHAPTER 9

Strategies

All the changes that have occurred in the several systems of ideas and resources in the village ultimately have to be explained by reference to the actions of specific individuals pursuing their own personal strategies.

A strategy is a selection of cultural information organized by an individual and implemented through action. There is no need to require that this organization be completely explicable by the individual or to require that it does not reflect personal habits, talents, tastes, and predilections.

The selection of information to assemble into strategies inevitably presents problems of weighting because of the incommensurability of the ideas of the different systems of information. The result is that the more valuable and important elements are generally given priority. If one has to choose between what is necessary for holding land and what is necessary for raising good goats, one sacrifices the happiness of the goats. As specific sets of weightings are constructed by villagers and seen to work, common patterns develop among households that use similar major resources. These patterns come to be generally known and themselves influence individuals, even as individuals shape the common pattern.

Consider one of the simpler possible choices: a villager evaluating the increasingly important activity of raising buffaloes. As part of knowing what buffaloes are in an ecological sense, he knows what they eat, what they produce, and what diseases they get. In the context of the division of labor, he should know what it takes to secure them. Economically, he knows their costs and yields. Obviously if he did not place some sort of positive evaluation on all this information, he would not consider buffaloes at all, but this evaluation is not the end of the consideration. Everyone knows that buffaloes do not exist only in one aspect or one context at a time. Their several conceptual aspects are brought together in fact because the buffalo is an animal in fact. Yet no one message source says how these several aspects will interact. What will all the separate pieces of information add up to together?

If a villager is wise enough to compete successfully, he will look carefully at the others who are raising buffaloes. Now things begin to get interesting, because he can immediately see consequences far beyond the ideas immediately associated with buffaloes as such in the several information systems. Buffaloes require food as an ecological matter. Economically and managerially, this in turn requires that the owner of the buffaloes should control land (rather than attempt to buy fodder itself). This in turn, again as an economic and managerial matter, means that one should also own land. In 1978 there was a linear (Pearson) correlation of .58 between the number of buffaloes a family owned and the amount of land it owned, and of .63 between the number of buffaloes and the amount of land from which income was obtained. These are trends of sufficient magnitude to be obvious to careful observers without benefit of the statistics themselves. But if one has land, one has to have the wherewithal to farm it, so there was, for example, a .60 correlation between the number of buffaloes owned and the number of plow animals owned. Buffaloes require much personal care, so a relatively large family is necessary (Pearson correlation of .46 between number of buffaloes and number of residents, as contrasted with .23 between the number of residents and the amount of land itself). Buffaloes also require dairy space, meaning (in 1978) a bricked floor where they can be tethered and preferably an enclosed barn. Finally, buffaloes require regular and dependable arrangements both for financing the herd and for marketing the milk products on a daily basis. Thus, when its implications are observed in practice, the seemingly isolated "tactical" decision to obtain buffaloes grows into a decision involving a long-enough time perspective and the marshaling of enough different resources to be called a strategy. The strategy isn't monolithic enough to define a person's, or a family's, entire set of social relations and resources, because, among other things, it leaves out religious preferences and a commitment to party activities, but any villager would know that one who raises buffaloes has several options in either area and nothing is fixed absolutely. Active party participation is likely because of the requirement for controlling land, but it may be possible to avoid participation if one prefers to. Religious identification is in principle irrelevant. In practice, in order to be understood by others one will probably use the ideas of the dominant (Sikh) religious system to explain any identification that one prefers, be it Sikh or non-Sikh.

Will the buffalo-raising strategy be adopted? Perhaps—if a person has a better balance of what is required for this than for other possible alternatives and also if it suits him. Other things being equal, some people may not like being tied to the daily routine that dairying requires, being prevented from moving about the region the way ordinary farmers (or nonfarmers) sometimes can for fairly extended periods. Others may see the routine as a thing valuable in itself. They may prefer to remain in the village for long periods and to keep their families together for work. They may even positively enjoy working with buffaloes. In most cases, the last factor in the balance between choices comes down to character and taste.

MAJOR STRATEGIES

Strategy considerations of this sort underlie and illuminate the occupational patterns of the different village groups described in Chapter 4. Not all resources can be quantified like buffaloes to show general trends, and no strategy is made up by manipulating only quantifiable resources. Social relations constructed with the moral and ethical concepts are also crucial.

Apart from buffaloes, major material resources include people, houses, other cattle, and land. With the exception of a few strategies unique to one or a few houses, the way these are handled divides villagers into two broad groups: those with farmland and those without it. For those with land, there were some slight differences between strategies aimed at maximizing total land owned and those aimed at maximizing farm income.

Strategies of Landholding

The previous chapters indicate several different ways in which the relationships among material resources relevant to holding and obtaining income from land changed between 1965 and 1978. How did these various changes interact and balance out? The answers are indicated in Table 9.1 and are based on an extensive correlational analysis of two main files. One represented the entire village in 1965, another in 1978. Additional information came from a third file reflecting those families who were represented in both studies, with a selected subset of their major resources at both times.

The first part of the table shows the main strategic correlations in 1965, the second part in 1978. As with the buffalo-dairying strategy, these coefficients themselves are of course not known in the village, but the relative em-

Table 9.1
Strategies of Land Ownership and Land Use, 1965 and 1978

	Family size	*Total residents*	*House units*	*New house units*	*Cattle*	*Plow*	*Land by income*	*Well land*	*Canal land*	*Land per capita*	*Land per resident*	*Income per resident*
1965												
Total land owned												
Pearson correlation	.410	.411	NT	NT	.412	.430	.870	.958	.961	.722	.643	.711
N =	78	78			43	33	78	78	7	47	47	47
Total Income												
Pearson correlation	.565	.581	NT	NT	.570	.503	.870	.850	.951	.552	.467	.728
N =	117	117			51	33	78	70	7	58	58	58
1978												
Total land owned												
Pearson correlation	.537	.384	.187	.003	.376	.081	.948	.883	.533	.616	.711	.644
N =	54	54	54	24	51	30	54	53	18	45	45	45
Total income												
Pearson correlation	.500	.368	.222	−.004	.465	.129	.948	.907	.305	.612	.706	.726
N =	62	62	62	28	59	32	54	53	18	51	51	51

Note: NT = not tabulated.

phases they reflect and the total number of households that combine any pair of elements are.

One of the most important points the table makes is related to professionalization. Basically, between 1965 and 1978 the labor pool represented by the number of family members in residence became relatively less important as a means of holding land, while income from nonresident family members became more important. More exactly, for landholding families in 1978, the number of family members outside the village was a more important determinant of the amount of land held and of the amount of income obtained from land than the number of family members in residence, where in 1965 the importance of resident and nonresident family members had been about equal.

This is indicated by the correlations between total land and family size, total land and number of residents, and total income and these same two variables. In 1965 there was a moderately strong correlation (.41) between the total size of landholding families and the amount of land they held and virtually the same correlation between the number of family residents and the amount of land (.41). This means that nonresidents and residents were both moderately important in obtaining and holding land; this also reflects the fact that there was a high correlation between total family size and total number of residents (.93). For obtaining income from land, rather than just owning it, the relationships were again about equal, but both were stronger, as shown by the substantially higher correlation coefficients (.57 and .58, respectively).

In 1978 the correlation between the total land owned by a family and total family size had increased substantially, to .54, while the correlation between total number of residents and total land owned had dropped to .38. This is what implies that the total number of *nonresidents* in a family was actually more important in determining the amount of land a family owned than the number of residents. As one would expect, the correlation between total family size and the number of family members in residence dropped correspondingly, to .84. At the same time, because the nonresidents did not require farm income for their maintenance but helped produce income from the land (for example, by providing cash to hire labor crews), the total family income from land had come to be slightly less strongly correlated with total family size than it had been in 1965 and even less strongly correlated with total numbers of family members in residence (.37).

Correlations within the file that combined 1965 and 1978 data showed the same thing in another way. The more members of a family working within the village in 1965, the less was its chance of having large amounts of land in 1978. There was a moderate negative correlation (−.27) between the amount of land owned in 1978 and the proportion of the families that had been residents in 1965. The negative correlation between the proportion of family members in residence in 1965 and the amount of crop income in 1978 was a little stronger, −.29. Thus a perceptive villager in 1978 reviewing the histories of other families could have seen that having family members outside the village had been important in determining the visible distribution of landholdings and would have to conclude that it was likely to be at least as important in the future.

Perhaps surprisingly, the total number of housing units was less important as a determinant of the size of holdings than family size. There are even suggestions that house construction was sacrificed during the 1965–1978 period so that villagers might invest in education and outside employment. This is indicated by the correlations between total land held and number of house units and total land and new units. They show only a slight tendency (.19) for families with larger amounts of land to have more housing units and no tendency for those with larger amounts to have added units since 1965. There was a somewhat less weak (.22) tendency for families with more income from land to have more housing units, larger houses, but again no tendency (or a very weak negative tendency) for those with more income to have more new units.

Traditionally, amounts of land were often stated in "plows," a convention that indicates a close relationship between holding land and owning draft animals. The changes in the ecology and in economic return from meat and milk production have made all animals, especially draft animals, less important as a means of holding land. This is shown by correlations between land and cattle and land and plows ("plows" here means draft animals only) and between crop income and cattle and crop income and plows. For 1965 the correlations had been moderately strong between all four relations, but strongest between the total number of cattle and the amounts of land farmed. Cattle included both draft and milk animals and this correlation reflected both the importance of cattle for power and for food for children, as well as the chronic fodder shortage (which kept families without land from having the milk cattle they wanted). In 1978 cattle were less important strategically as a concomitant of landholding and farming, but the most spectacular drop was in the connection between plow animals and the amount of land owned. This relationship went from moderate (.43) to no relation (.08). At the same time, the relation between plows and actual farming income dropped almost the same amount, from moderately strong (.50) to weak (.13).

I have already said that in 1978 landownership had become almost the only basis for using land to obtain income from it. The converse of this is that farming one's land was the usual way to hold it. This is indicated statistically by the increased correlation between total land owned and total farm income, from .87 in 1965 to .95 in 1978.

In addition to deciding what to combine with land, farmers also need to decide whether or how to combine the various types of land with each other. The changing strategies for balancing land types with regard to what they can best produce was described in Chapter 4, but there are also questions of scale. Should larger farms have different mixes of land just because they are larger? Similarly, should farmers attempting to optimize local income have different mixes than those trying to leave the village? Miscalculations in such matters are very difficult to remedy by minor adjustments. The overall answer was that in 1965 larger farmers tended to have more canal land, but this broke down in 1978, while the proportions of different types of land did not change systematically with total farm size in either year.

Table 9.1 shows that in 1965 there was a very high correlation (.96) between the total land held and the amount of canal land held. The correlation between canal land owned and the total family *income* was slightly less (suggesting that some canal land was not used for farming by the owning families, but was let on rent or mortgage). The correlation between total land held and the amount of well land held was about the same. In 1978 the correlations between the total amount of land and the amount of canal land a family owned had dropped to only moderate (.55) and the correlation between total income and the amount of the canal land owned had gone to weak (.31).

The main reason larger farms in 1965 tended so strongly to have more canal land (and why this tendency weakened in 1978) was just that canal land was rare. Only a few farmers had it and these tended to be larger ones. When the proportions of land of each type are correlated with total size of holdings, the relationships are either very weak or nonexistent. The highest correlation in 1965 was between the proportion of canal land owned and the total amount of land farmed, but the correlation itself was barely moderate (.26). The next-highest correlation was between the proportion of well land owned and total land farmed and it should be described as low (.19). All other correlations were substantially less, though positive. In 1978 the relations between total sizes of holdings and the proportions of land of the different types were similar, but weaker.

In 1965 canal land had two major advantages. No fodder was required to water it, and it could be rented out by nonfarming owners without fear that it would be damaged. By 1978 the two advantages had been reduced to one. Since the fodder shortage had been relieved, canal land was no longer needed by resident farmers to supplement fodder production. Hence its advantage was mainly as an investment safe from poor maintenance. Well land was relatively more desirable for farming because it was cheaper, more available, and sufficiently productive of fodder to take care of itself.

The final group of correlations in the table pertains to the drive for more land and the tendency for operational holdings to become larger, a tendency that runs through all of what has been described. It is obvious that land is important, but why always struggle for *more*? Some have argued that it is simply an irrational Jat predilection, like the supposed inability to save. The table suggests a better reason. In both periods, it appears that as farms became larger, both the amounts of land owned per capita and the income per capita increased.

In 1965 there was a strong tendency for families that owned larger amounts of land also to own larger amounts per family member (.72) and per resident member (.64). That is, wealthier families were also wealthier per capita, and income per resident rose strongly with total income (.73).

In 1978 the general tendency for land owned per capita to increase with total land ownership was virtually unchanged, but because of the much tighter relationship between the amount of land owned and the amount of land farmed, the correlations between total farm income and total land per capita as well as total land per resident both increased substantially. The strategy of landownership obviously takes on the character of a rather fully defined way of life once its implications are revealed by examining the way actual families combine elements from the several systems into common general patterns.

If you want to farm, you have to be prepared for a continuous effort to get more land and this means that (by 1978 especially) you have to get more money to obtain it by purchase. In order to do this, it is necessary to exercise far-reaching control over family affairs in a most difficult way: you have to make sure that a substantial portion of your family members have the capacity to work outside the village but will not use this capacity to cut themselves off from the remaining members and stop cooperating. This is directly related to the concern over kinship concepts and equally directly related to the widespread support among farmers for the creation of the local school—not indeed as a side interest but as part of their pursuit of basic household strategies. Some of the interest in supporting the local religious institutions may have had the same source. It is certainly easy to see how the Sikh ideas of self-sacrifice, the importance of living a working life well, and religious tolerance, all would appear especially valuable to those responsible for managing families in this type of strategic environment. Finally, while a farmer will obviously need a house, he cannot count on being able to increase housing, but indeed may have to forgo housing improvements in favor of the more basic aims of land acquisition and education or vocational advancement. The statistics do not show it, but from what has been said it should also be clear that a farmer should expect to spend a good deal of his time outside the village on various errands and in handling financial dealings.

Strategies of Nonlanded Families

Families who are not farming naturally show quite different ways of combining people, cattle, and housing. Table 9.2 gives the correlations among population, cattle, and housing for those who do not own land and compares them with correlations among the same resources for landholders. Estimates of size of houses and new house construction were not coded into the 1965 computer study and had not been collected as systematically as in 1978. Therefore, the table reflects 1978 only.

The most striking general point is that the landless families show substantially higher correlations between total family size and both the number of residents and the number of men. This contrasts with lower correlations between most demographic variables and the indicators of housing size: number of units and number of rooms. Evidently the main strategy for most landless heads of families is to keep their family groups together and in the village. It is thus less important (or less possible) to match housing facilities to family needs. The gap is not great, however, and the correlations are quite close between the total number of rooms and family size and, perhaps more important from the point of view of provision of separable areas for families with several sons who have wives, between the total number of rooms and the number of adult males (.43 for landless, .54 for landowners).

Among the landless, the correlations between demographic variables and new housing construction are all substantially weaker than among the landed and are insignificant. Larger families were, if anything, slightly less likely to have had new construction than smaller families. Among landholding families there was a moderate tendency for the reverse to be true. In both cases, however, it should be noted that the strongest correlation was between the number of new or remodeled units and the total number of units—.65 for landless families and .86 for landowners. That is, regardless of family size, there was a general tendency for those with houses to improve them. After this, for landowners, there was an almost equally strong correlation between new construction and total number of "plows," .83. This suggests or reflects two things. First, among active farmers, a high priority in the construction of new "units" (as opposed to "rooms") has been for the provision of new yards for tethering cattle. Second, for all groups, new construction has, as previously noted, most often been associated with divisions of families—splitting up, building a new unit and then, often, dividing the total housing on some equitable basis.

It should not be forgotten that between 1965 and 1978 a significant number of families chose to abandon farming. That is, a strategy choice developed that consisted in diverting resources from landholding and marginal

Table 9.2

Strategies of Landed and Nonlanded Families, 1978

						Nonlanded families		
	Family size	*Total residents*	*Total men*	*Total cattle*	*Plows*	*House units*	*New units*	*Total rooms*
Family size		.91 (118)	.79 (118)	.37 (70)	.89 (4)	.34 (104)	−.50 (43)	.34 (89)
Total residents	.81 (69)		.72 (118)	.43 (70)	.91 (4)	.24 (104)	−.17 (43)	.25 (89)
Total men	.88 (69)	.70 (69)		.35 (70)	.69 (4)	.43 (104)	.10 (43)	.43 (89)
Total cattle	.37 (51)	.65 (51)	.28 (51)		.58 (4)	.09 (70)	.08 (34)	.40 (59)
Plows	.05 (30)	.14 (30)	.22 (30)	.44 (30)		NV	NV	1.00 (3)
House units	.33 (54)	.38 (54)	.34 (54)	.29 (51)	.59 (30)		.65 (43)	.52 (89)
New units	.23 (24)	.11 (24)	.28 (24)	.31 (24)	.83 (14)	.86 (24)		.40 (41)
Total rooms	.48 (51)	.43 (51)	.54 (51)	.34 (49)	.53 (29)	.53 (51)	.55 (24)	
		Landed families						

Note: Coefficients above the diagonal reflect all families with no land. Coefficients below the line reflect the same parameters for all families with land. In each group of numbers, the upper is the Pearson linear correlation coefficient, the lower is the number of households having both the variables. NV = no value.

farming on insecure arrangements to seeking wage work in a more professional environment. The discussions of Tables 9.1 and 9.2, taken together, pertain directly to this choice.

First, it should be noted that the relation between increasing size of holding and increasing per capita income would affect small farmers two ways: they would realize that they would do better if they obtained more land and they would be under constant pressure from larger holders to sell out. To buy land, a smallholder would have to get cash. This in its own turn would mean that, like larger holders, he would be under pressure to send one or more members of his family off to work. But for a small farmer, with less income per capita, supporting a search for employment is more of a burden. In addition, earned income for a member of a small farmer's family is likely to be lower than it would be for an equally capable member of a larger family, because it is highly likely that he will not have been able to obtain as much education. Thus the choice could be this: "Shall I continue to struggle to hold land when in all probability I will in the end be unable to improve my standing among landholders, while my sons (and possibly daughters) will be qualified for only the lowest type of job? Or shall I pull out of farming and concentrate on keeping my family working in and near the village while they obtain more education, so that in a few more years they will be able to get much better jobs and move out—or even work in an office or school nearby and stay in the village at an improved level of welfare?" The real long-term gains, even seen strictly from the point of view of the head of the house alone, could easily be greater with the latter course—depending, once again, on the character, talents, inclinations, and basic interests of the people involved.

With this said, it would belabor the obvious to explain why the Harijans did not complain about the school's encroachment on "their" cremation grounds, or why the school had the support of Potter and Goldsmith panchayat members—as well as that of virtually every other occupational group in the village, in addition to the landowners.

Within the broad general strategy for nonlandowners, separate courses of action depend very largely on particular trades that are followed which require either the possession of special training (certification as a teacher, for example) or ownership and use of specialized equipment (like looms). The numbers of families in each such occupational category are too small for a correlational analysis to add anything to what has been said in Chapter 4.

FOUR HOUSEHOLDS

There is more to life than statistics—even more to life than material success. General trends indicate patterns of choices, but they do not predetermine them. A rational person will always take the trends into account but will not follow them blindly. There is always a moment of choice when one has to ask not only what the trends are but what they promise personally. The answer will differ from one individual to another, depending not only on their resources but, as already said many times, their character, habits, and tastes.

Villagers know each other intimately. The great majority of the men spend their entire lives in each other's company, like it or not. Yet circumstances forbid them from helping one another every time they see a need, resolving conflicts every time they arise, or even avoiding conflicts every time interests clash. So beside the intimacy there is detachment. In this spirit they usually describe their own and each other's careers and purposes. In the same spirit I will try to tell the stories of four villagers. The first of these, following the rough order of *jātīs* of the division of labor, is Babu Ram.

Babu Ram

One does not have to be in India long to learn that the normal mode for doing business with shopkeepers is what Americans usually call "haggling." Locally it is more often described as "keeping everything open and above board." I became accustomed to this before I settled in the village, so when I went to Babu Ram's small shop to see what he had and later to ask questions, his behavior and the way others dealt with him struck me as quite unusual. Transactions always occurred quietly and with no argument. Babu Ram stated a price and it was paid with clear deference on both sides. Since I could not ask Babu Ram "Why are you respected?" or "Why are you polite?" others explained. Babu Ram was an honest shopkeeper.

I was told that when Babu Ram had come to the village about ten years before, replacing a relative who had left, everyone naturally assumed that he was cheating. So when he stated any given price they always insisted on paying less. After two or three years of this, the farmers one day saw Babu Ram packing up, very sadly. They asked him where he was going and he replied that he was leaving the village. They asked why. He said he had no more money and could no longer keep the shop. It was then that they realized what had been happening. Babu Ram had been stating fair prices all along, but had been accepting what was less than fair. Several farmers set him back up in business and he had stayed. He cheated no one and no one cheated him.

Babu Ram was too self-effacing to be very talkative. One of the first of my more extended conversations with him occurred about three weeks after I started work. Caste rankings were considered important by American Indianists at the time, so I thought I should do at least a small survey, if only to allow comparison with other areas. I therefore obtained a list of *jātīs* in the region from several villagers on several separate occasions and wrote their names on 3″ by 5″ slips, in Gurumukhi. After a few pretests on those who had given me the names, to see if the operation of ranking itself made sense, I set out to find a sample of new people of the various castes. Babu Ram was one of the first I asked, since he was a Brahmin and since I judged from his behavior up to that point that I could explain the project to him and he would give a candid response.

Two farmers were sitting on the burlap sacks that lined the concrete floor of his shop when I arrived. As I entered he asked me to sit as well, and I took a place beside them. I explained to Babu Ram the purpose of the cards and that I would like to have him indicate the rank that the named *jātīs* had in the region—if he agreed it was a clear question. He asked if I truly wanted his opinion. I assured him that I did, his own opinion. He took the slips and read through the stack carefully, moving each to the bottom after he read it. When he had read them all he asked again about the procedure. I again assured him that I would like him to indicate the rank of the *jātīs* in the region. He then said, quite deliberately (but, I thought, a little wistfully), "Well, the Jats are highest, because they own the land." He pulled the slip for the Jat *jātī* out of the stack to move it to the top. The two farmers, Jats of course, immediately protested. No, they said, the Brahmins were on top, because they were most respectable. Babu Ram looked pleased in a very quiet way, but repeated: "No, it must be the Jats. They own the land. All things come from the land." The farmers again disagreed. This continued for several minutes. I kept silent and tried to be invisible. Babu Ram would not put his own *jātī* on top. Finally, one of the farmers, raising his voice, half-rising and showing considerable agitation, commanded: "The Brahmins are on top! You put the Brahmins on top!" Babu Ram finally, silently, complied, thus actually confirming the superiority of the Jats and persisting in his refusal to show pride, although it was not hard to see that he was pleased (again quietly) by the compliment. The Jats came next in rank.

In 1965 Babu Ram gave his age as 57. He lived with his wife aged 45, two daughters aged 9 and 12 and one son—Dharmpal, aged 6. His wife had borne three other daughters, but they had not lived. The loss of the girls was very hard on Babu Ram and very discouraging, but he was pressed by the need for at least one son—not only to care for him but also, of course, for his wife's sake. After Dharmpal was born, Babu Ram personally took care of him and refused to let his feet touch the ground for the first year.

On some rituals when it was customary to take a gift to the hereditary village Brahmin, families would take one to Babu Ram. The idea was not only to express respect, but to say "You are here; you are part of this village, as much a part as anyone can ever be."

Babu Ram's shop was in two long, narrow rooms opening off the main street, in a group of old brick buildings that had once been village offices. He lived in these same two rooms and also used part of a very old farmhouse next door, mainly an upper-story "guest room," ancestral property of Indar Singh. He sat waiting for business on the concrete floor of the shop every day and would always open the shop if anyone came by at odd hours. He sold a few basic school supplies, some of the least-expensive spices, cigarettes, kerosene, vegetable cooking fat, and salt, which he brought from nearby towns on his cycle. He also bought and resold some local vegetables and staples, such as onions and potatoes. He had no cattle, but he did own about an eighth of a *bīghā* of rainfall land (about a hundred square meters), which he rented out. He would buy the ghee and milk he needed locally. Around January 1966, as I recall, he began construction of his new house on a plot of land just inside the village gate. I was told this land was "given" to him by a farmer who had previously refused to sell it to the village doctor.

Babu Ram's elder daughter was married in 1971, the younger in 1976. In 1978 his age was given as 66, Dharmpal was 17, and there was a second son aged 10. All were living in the new house. Babu Ram had "retired" a few years before my return, due to ill health (a topic no one seemed to want to dwell on). He had retained the store, but the roof had partly fallen in, the door was broken, and it was used only for storing straw. He owned no more land, but did have a buffalo. As mentioned at the outset, Dharmpal had finished high school and was awaiting word from the employment exchange in Chandigarh, where he hoped for a job as a "servant" (which I took to mean low-level government office employment).

When I first returned to the village and wanted to see Babu Ram, I was told he was ill and was asleep. Two weeks later, I stopped by again and was told the same. Finally when time was getting short, I explained I would be leaving soon and would come back that afternoon, which I did. I was bid in and found him apparently just waking on a cot inside his door. He had allowed his beard to grow out and was wearing a *kurta* in a short-sleeved pattern in place of the long-sleeved, lighter-colored, and more traditional type he had always worn. As for illness, the most apparent problem was starvation. He had never been robust; now he was thin and his movement was delicate even by his own standards. The thought immediately came to me that since he was not earning, he had cut his consumption to the absolute minimum—probably to less than that. This of course could not be said or asked about, nor could anything be offered that he could accept. Apart from his physical condition, nothing had changed. We spoke, but it was difficult to say very much. It was painfully obvious why he was not working then, and it surely would not have been anything but illness that stopped him from working before. There was nothing to ask about the past and the present was best noted in silence or indirectly. It was too evident that the illness had set off a tragic cycle and all that anyone could do was stand by and watch, or help in only the subtlest of ways. It was going to be up to Dharmpal.

Was Babu Ram following a coherent strategy? Despite his difficulties, the answer would have to be yes. Did it follow the general trend for landless families? Yes. Is that why he did it? No. What else could he do but bend all efforts toward having at least one son and educating that son to the point where he might obtain a job good enough to replace the shop? Sixty-six is not especially old in the village, but no one can keep working forever. Moreover, the strategy appeared to be paying off. Was Babu Ram being unreasonably proud to refuse help and rely on himself as he had? In his own terms, he was not proud at all; he would refuse help not out of his own sense of honor so much as out of a sense of respect for others and a desire to avoid imposing on them any obligation that would deprive them of doing all they could for their own children. The respect he earned by this behavior was unquestionably an asset in the village, but most important must be reckoned the respect of Dharmpal himself—both for himself and for his parents. He was a fine young man by everyone's account, so far as I could see, and it is highly likely that Babu Ram's example and the place he earned in the village had much to do with this.

Sher Singh

A quite different but equally coherent set of strategies is presented by Indar Singh's immediate neighbors, the family of Sher Singh. Sher Singh is of Jat *jātī*, of a different *pattī* from Indar. In 1965 the family supported the dominant party, the same as Indar Singh. There were fifteen people in the family, Sher and his wife, four sons and their wives, one unmarried son, a son and daughter of the elder married son, and a son and daughter of the second eldest. They owned two oxen, one ox calf, three buffaloes, one female buffalo calf, and one camel.

The family then owned about 41 *bīghās* (3.5 hectares), of which 61 percent was well irrigated, 15 percent was well-canal land, and 24 percent was rainfall land. This gave them substantially less than the average amount of land per capita and they were trying hard to add to it. By mortgage and tenancy arrangements within the village, they had increased their farm income to the production equivalent of almost 53 *bīghās*, mainly by taking additional well land and rainfall land (letting 16 percent of the well-canal land). But more important in the long run, the family was working to obtain new land in other areas. The youngest married son and his wife had moved to a new farm the family was just buying in a village about thirty kilometers away, in Ludhiana district. The second eldest son, Master Gurdev Singh, divided his time between the two farms and in addition spent a great deal of time traveling in the area.

Gurdev Singh was a high school matriculate and had earned an additional certificate for primary schoolteaching. The title "Master" reflects the student practice of addressing teachers as "Masterji." He was a large, handsome man, then about thirty years old, with a manner at once genial and forceful. He spoke good English and was knowledgeable and sophisticated about government and practical matters of many sorts. It was he who first explained to me the effect of the Security of Land Tenures Act on the regional market in land and the system by which villagers (including most families in his *pattī*) were responding to the opportunities. My impression is that many of his activities outside the village involved helping friends and relations secure such land, as well as helping villagers fight their various court cases. He was the *dādā* of the headmaster of the nearby school previously noted and he was an important ally and supporter of the headmaster in regional political matters—and of course in the village itself. Sher Singh himself was of a quieter disposition, but there was no question that the family was united and that Sher, and not just Gurdev, was responsible for the general strategies.

In 1978 the total family size had increased to seventeen, although the number of residents had been reduced to ten. Four of the nonresidents were on the new farm in Ludhiana (the original couple and their son and daughter), two were daughters of Gurdev and his elder brother who had recently married and moved out—one to Chandigarh, one to another village in Ludhiana district where her husband was engaged in farming and dairying.

In the clash over the elections in 1972 and with the headmaster over the school, Sher Singh and Gurdev had stuck with the headmaster and with Siani Singh in the village, splitting off from the dominant party and becoming major figures in the new minority party.

In 1978 the men of Sher Singh's family were not on speaking terms with the men of Indar Singh's. I'm reasonably sure the conflict was being kept alive by current land and political disputes, since none of the men in either family were the sort to be hostile simply out of habit or without cause.

The new farm was well established and apparently about the same size as the operation in Shahidpur, although I have no hard figures. In Shahidpur itself, the total land owned had increased by a half-*bīghā*, but the family had stopped farming land on bases other than ownership. What it farmed was what it owned and all the land owned was farmed. Fifty-nine percent of this was well land, 14 percent was canal land, 18 percent was well-canal land, .08 percent was rainfall land, and the rest was work areas, well sites, and the like. Of course the wells were all mechanized, so the water capacity was many times what it had been in 1965. The family had acquired a 5 HP diesel motor for use in Shahidpur and an electric motor for the new farm. Finally and most important, they had bought a large tractor, which was used on both farms and hired out on a contract basis. In Shahidpur, the family owned two oxen, four buffaloes, two female buffalo calves, and one small goat.

In 1965 the family had lived in a comfortable but modest house of five rooms, of about 100 square meters in floor area, all of brick construction, with a pleasant upper-floor guest room. A second, older, unit of four rooms was used for cattle and work areas. In 1978 the original two units were the same, although the main house was more nicely painted, but two more had been added: an open field within the *abādī* as a work area and a field with a two-room building on the edge of the *abādī* for the tractor and the cattle. Although the increase in housing was mainly in the way of increasing its value as agricultural capital, there were still clear touches of relative luxury: two electric fans, two radios, three bicycles, and chinaware and glassware replacing the more-durable brassware for eating and drinking utensils.

Shortly after being told of the party realignment and that Gurdev would therefore not come to Indar Singh's house, I stopped by his own place in the hope of finding him (and to avoid associating only with one group). He was not there then, or later, but I had tea with his brother and although we had not known each other well during my earlier study, we reviewed the general progress. He described it in the same strong terms as the group in Indar Singh's house and I am sure he was speaking for his own family as well as for the village and region at large.

Has this family been following an efficient, effective strategy? Different though it naturally was from that of Babu Ram, the answer must again be yes. The expansion of their operations and their apparent increases in income and welfare at least equal and probably exceed the general village trends. The family has advanced its position economically and maintained its prominence in village and regional affairs. Does the strategy accord with what the general trends for landowners suggest? Yes, although seeking land outside the village is a new refinement. Have its members been "purely" calculating and manipulative? Of course not. Sher and Gurdev's support of their friends and relations is heartfelt and based on genuine ethical convictions. The break in their relations with Indar Singh's family was a source of emotional discomfort on both sides. But only the callowest sort of analysis would say that simply because convictions are deep, they could not motivate patterns of effective and purposeful action. A good part of the success of any family depends on the extent to which others will trust its members and consider them reliable in meeting their responsibilities.

Indar Singh

The strategy of Indar Singh's family differs from that of Sher Singh mainly in not involving a shift to farming operations outside the village.

In 1965 the head of the household was Bakhtavar Singh, Indar Singh's *bāpū* ("father") and the only man in the village who was a *paRdādā*. He was at least seventy-five years old, but lively and in good health. He always took meals at the well, where the family had a brick house used for storage and for keeping chickens, rather than in the house in the *abādī*. He also slept there—all year, so far as I know. When the family was working, he insisted on doing his share, rotating in to do what he could if hands were needed and rotating back out to rest if stronger members were there to take his place. We could not speak much. Being unable to read, he was not accustomed to breaking his speech into words and slowing it enough for me to understand. My speech likewise was difficult for him. After several months in the village, I naturally asked why he never came to the house. The explanation was that it was because of "respect." Because of the deference he would be showed (in part because there were only people much younger than himself left alive), it would be difficult for those in the house to go about their business. So to avoid inconveniencing them he had retired years before to the well.

Indar Singh was Bakhtavar Singh's elder son, fifty-four at the time. He had one younger brother, Santokh. Santokh had worked on the farm while Indar served in the Army, but as in many Panjab farm families with holdings too small to divide into two viable units, he had never married. Indar Singh had married twice and had four sons and one daughter. The eldest was Ajmer, whose age was given as twenty-three. Netar, the next eldest, was then eighteen. The first wife had been their mother. After she died, Indar Singh married her younger sister, with a new ceremony. She was the mother of Netar Kaur, Indar's only daughter, then eleven, and Surindar Singh and Darshan Singh, nine and four, respectively. Ajmer had left school early to help with the farming and had one son, the same age as Darshan. Ajmer's wife had died in a household accident, perhaps two years before. Netar had been engaged to her younger sister, again following a common pattern, but because her family was not certain of the circumstances of her death, they had stopped moving ahead with the marriage arrangements. When Netar explained this early in 1965 (after asking about marriage customs in the United States), he had to conclude that he did not then know if he was married or not. Netar was then in his last year of higher secondary school. The younger children were also all in school, according to their ages.

The family's income was mainly from farming, although Indar Singh had a military retirement pension. They owned a little over fifty-one *bīghās* of land and farmed the equivalent of forty-eight, on the record, of which about thirty-eight were well land and eight were canal, of good quality but not extraordinarily so. In addition, they took care of some of the land of Hari Singh, which borders their own, and rented some from another relation who had moved out to another village. Hari Singh was away serving as Subedar, a class of Indian commissioned officer, in the Army. His father's father's father was brother to Bakhtavar Singh's father. The family had one well of their own and also the use of another owned by Hari Singh. Their cattle included one camel, two oxen, one cow, one ox calf, and two buffaloes. Their house was about the same size as Sher Singh's and had been built by Indar Singh, he said, with his own hands—apparently shortly after his military retirement. In addition to it, they still retained their former house next to Hari Singh's (where Babu Ram had his room) and had three other empty plots within the *abādī* used for various work purposes.

The family was moderately active in party matters supporting the dominant party (of Siani Singh), apparently mainly because of their involvement in helping Hari Singh arrange a wedding for his daughter in June 1965 and because of a conflict over one of the work areas.

Netar Singh graduated high school and began college in 1965/66, while I was still in Panjab. He chose a college that was near his mother's natal village and stayed with her family while he attended (growing and filling out rather spectacularly because, as he explained, they had an extra buffalo giving milk). Eventually, after graduation, his marriage was completed and a daughter was born in 1974. A second daughter was born in July 1978. As noted, he was working as an agricultural grain inspector for Markfed in a village about 200 kilometers away, but was most anxious to be transferred nearer home. It was considered good work, although by no means easy, and encouraged the younger members of the family to press on with their own schooling. Surindar, also as noted, was working on his Master's degree in English History. He had completed the first year and was awaiting the examination results when I returned to the village.

Bakhtavar Singh's death occurred not long before my return. The family had undertaken all the funeral arrangements carefully, in the traditional manner, and Netar was insistent that I record them, because "now, as education is increasing, these things are becoming forgotten."

Indar Singh had always been an innovative farmer, looking for new ideas to try but proceeding very systematically with his assessments in the initial phases. In 1965, for example, when my wife was coming to join me, I asked her to bring some packets of American sweet corn. The Indian maize was field corn and not good for eating steamed, on the cob. I was curious to see if I could do better. Since it was impossible to get land in the village, I put some of the seed in large clay pots on our veranda in Chandigarh in about March of that year. The plants grew to two or three feet in height, but in May or June they withered and died from the heat just as the ears began to form. It happened that Netar came to visit before they were disposed of. He asked about them and I explained. American maize enjoyed an excellent reputation in Panjab and I suspected from his reaction that Netar thought the problem was more likely to be with the farmer than with the crop, but he did not say so. He asked if he could have some of the seeds. I gave him the remains of the packet and thought no more about it.

Late the following September or early October, Indar Singh and I were walking out toward his well. He asked if I remembered the seeds. Yes, certainly—what did they do? He replied that I had given him ninety-four seeds. He kept eighty-four and gave ten to his wife's brother, who was a very good farmer, to try in his own village. He planted all of his seeds in a plot by his well, the best land. His wife's brother did the same. When we arrived at the plot, he showed me the results: seven dry, stunted plants, the tallest barely sixteen centimeters high. His wife's brother's efforts yielded no plants at all. The experiment was appropriate to the problem and its result was clear. The variety was not well adapted to local conditions. In a similar spirit and with similar care, Indar Singh had been one of three farmers in the village to accept the state government's recommendation to raise chickens for meat and eggs. I believe he had about a hundred birds, including three cocks. The venture had done fairly well until early in 1965, when the price of feed rose and he had to abandon the project. Finally, still in the initial period, Indar Singh was one of the first farmers in the village to hire the Mason to set up and run an apparatus for mechanized threshing of wheat (for the reasoning, see Leaf 1972:141).

After 1966 the family kept abreast of developments and apparently tried all or nearly all of the innovations as they became available, except for dairying. They went through the sequence of new wheat varieties that were recommended, keeping those that worked best. They subscribed to the magazine of the Panjab Agricultural University (being one of two families in the village that the postman said received it in 1978) and they also tried, as before, to experiment with varieties apart from those that had caught the government's attention. In 1975 they were among the first two or three families to begin growing okra for seed. They also began to grow a kind of melon, called *peTī*, as a commercial vegetable. The suggestion for okra came through a friend in Sher Singh's *pattī* who had a relative who was a wholesale seed distributor in Delhi. The distributor provided the seed and promised to buy the crop, an arrangement which continued up to 1978. I do not know where the idea for *peTī* came from, but the crop was apparently marketed locally. Additional new crops by 1978 were, as noted, eggplant and rice. The family's decision regarding rice is especially interesting and reflects the development of the water supply in a way that is slightly more subtle than what has been described so far.

Indar Singh was one of the first in the village to adopt electric well pump motors after electricity came in 1970. At first he used a tube bore, but had a great deal of difficulty with the filter screens rusting out. Eventually he found an arrangement for mounting the motor with a pump in a pit about two and a half meters deep next to his old shaft well, while still using belts from the motor to drive the new machinery on the surface. The electric rates were charged on a flat-rate basis, as noted, but initially the rates were low. The water supply assured the winter crop against loss from too little water, but it could not help with the opposite problem in the summer: too much water. Winter crops were doing very well, but when the summer rains were heavy, it was still possible for the family (and farmers generally, of course) not only to fail to make an equivalent profit, but to lose its investment entirely. This is where the advantage of rice became most apparent. It was the one crop that floods could not damage. At the same time, the electric pump could assure a reasonably priced supply of water in the summer if rice was in the fields and the monsoon failed. The first rice had been planted by the family in 1977; 1978 was the second crop. A letter from Surindar in 1980 indicated that the strategy had worked consistently thereafter.

According to the 1974 land records, the family owned about three *bīghās* less land than it had in 1965, a result of some minor adjustments. The amount of well land increased by about two *bīghās*, the well-canal land increased by four, and the canal land decreased from eight *bīghās* to none. Total land farmed, just over fifty-three *bīghās*, was more than this and more than that farmed in 1965. The increase was mainly canal land held on mortgage, I believe from Hari Singh.

In 1978 Indar Singh said he was paying taxes on seventy-two *bīghās*. My impression was that most of the increase represented short-term arrangements like mortgage and tenancy mainly with Hari Singh, who was still away.

The family's housing in 1978 was the same as in 1965, except that they had half-interest in a new walled plot of land next to the village gurudwara. The other half was held by Hari Singh, who used the plot for a small garage for his automobile. Another portion of it was covered with a brick floor and a friend of the family (and brother of Naranjan) used it for a buffalo dairy. The remainder was a work area, with a few trees. The main house itself had not been improved. In fact, heavy monsoon rains a few summers before had caused the foundation to shift and a large and dangerous looking crack to appear in a wall near the front gate. Indar Singh was well aware of the need for repairs, but said that investing in the land had been placed first. So too had investment in education. The family had also been saving for the marriage of Netar Kaur, which was to take place in a few months. Of course, there were the touches of relative luxury I have already remarked on: the backed chair, electric lighting, several sets of china tea service, glass bowls for eating in place of brassware, the table fan, and a radio (of sorts—a battered portable whose own speaker had evidently passed on, wired into a large public address speaker, which it could barely power). But the amounts spent on these were very small compared to house repairs, well equipment, buying new land, or education.

Again the question: has Indar Singh's family been following a coherent, efficient strategy? Short of subjecting his detailed cropping patterns to a modern cost-benefit or linear programming analysis, it would be hard to find a set of actions that better deserves the label. If the management system that makes the best use of what is available at each point in time is also the one that makes the best use overall, it is difficult to imagine anyone who proceeds more systematically than Indar Singh to make sure that each decision he makes is in fact optimal when he makes it.

But does this mean that Indar Singh is some sort of "economic man," who calculates everything in terms of its material yield and ignores "social" values? Certainly not, because such a simple either/or choice is not part of the situation. Almost everything has a price and can be assessed on that basis, but almost nothing can be assessed solely on the basis of price. The question of which crop will grow and how well is separate from the question of the profitability of the crops, and the point of Indar Singh's crop experiments was to assess each of these two facets in their own proper contexts. But even this does not cover the whole range of considerations. How does one measure the return on educating one's children compared with buying more land? How does one measure the return on marrying one's daughter well, as compared with repairing one's house? One can estimate economic costs and returns in either case, but I am sure that for Indar Singh such an estimate cannot in any way summarize the matter. Rather, the result of the estimate has to be measured against other values, deriving from other systems. Education is, in Indar Singh's opinion, a good thing in its own right, as is preparing one's children to do as well as possible in life. Indar Singh also takes very seriously the general obligation of men, as owners of property, to provide for women as dependents and especially the obligation to provide good marriages. Once, in 1965, he described the tensions that were created when an affinal relative had declined to spend the money necessary to pay for his own daughter's marriage and Indar Singh had stepped in and assumed his obligations, simply because it was needed.

If one honors one's obligations, one can then appeal to others to honor theirs; one earns respect. But this too is not all there is to the matter. In the last analysis one reaches, for any major decision, a point where one can calculate no more. One can see, perhaps, that economic and moral considerations all tend to lead in the same direction or that they do not. Either way, some balance has to be struck. At that point, one consults one's priorities.

I hesitate to mention the most dramatic illustration of this. It was related during one of those personal conversations that I didn't consider putting in my notes, so I may have the chronological details tangled, but about in 1973 Indar Singh suddenly fell down, literally, with fits and spasms—unconscious and, as I believe he said, foaming at the mouth. He was, he said, completely insensible and stayed that way for weeks while the family took him around for diagnosis. Eventually, although free care was available at the government hospital in Chandigarh, the family decided to take him to a very well-known and widely respected missionary hospital in Ludhiana, because the care was better. He was operated on, a tumor was removed, and he returned home—in about ten days, if I recall correctly. He removed his turban to show me his head. The upper part of the cranium was very irregular and slightly scarred, but otherwise fine. The cost had been Rs. 5,000. I doubt that there would have been the means to make such an expenditure in 1965; in 1973 the money was there. Was the family going to calculate the return? What was it? The purpose of any strategy is to preserve and if possible improve life. The notions of what makes a good life and what sustains it therefore frame the other calculations and not the reverse.

Baldip Singh

Although enough has been said to illustrate the nature of strategies and some successful options, a brief illustration of the other side of the coin will make the picture more complete.

Baldip Singh was the minority member of the panchayat who had been frustrated in the conflict over the watch. He had been very helpful in the first weeks of my study in October 1964, but after this incident he became quite distant. It was not until I was writing this that I realized he must have decided I was secretly supporting the dominant group.

When I first met him, he had not been long returned from his service in the Indian Army. This had included a tour in Bhutan, then a frontline area facing China. He was a careful, disciplined, and progressive farmer. He spent long hours in his fields, where I would join him to ask about the work. He was the first to spell out the complex reasons that farmers have for doing almost everything they do. He was also the one who explained to me the system for calculating yield to predict annual income used in Tables 5.1 and 5.2. He had a modest house of three rooms and an additional unit of one room. He was then about thirty years old, of about average size but with an air of exceptional health and strength. His family consisted of his mother, aged fifty-five, his wife, twenty-eight, and three sons, aged nine, seven, and four. He owned about thirty-eight *bīghās*, of which about thirty were good well land and just under eight were canal land. He farmed just under twenty-five *bīghās* of well land and just over six of canal land. This was a very substantial amount for one man alone to care for, so he had hired a full-time servant to live in the house and help. For draft power, he had a very good camel and two oxen which he kept in his second unit. For the family he had two buffaloes which were kept in a room in the main house. He described his holding as small and was deliberately trying to make up for this by farming more intensively than usual.

When I returned in 1978, I did not see Baldip Singh in the village. He did not come to Indar Singh's house. I did not see him in the streets and could not find him in his fields, although I went several times. When the census was completed and I compared his old and new forms, they showed that his housing in 1978 was unchanged and that only a diesel engine had been added to his tool inventory. In cattle, he had lost draft power but had not added milk animals. He had had two oxen, two buffaloes, and two buffalo calves. When the copy of the new land record finally became available, initial inspection suggested that he had lost ground substantially. I was very interested in knowing how these losses had come about, but was reluctant to go to Baldip's house because of his previous desire to avoid my questions. Finally, when a small group was sitting in Indar Singh's house and I was asking his wife about some of the names on the voter registration list that the men could not account for, his name came up and I mentioned my impressions. Were they correct? The answer was yes. How did it happen? I don't remember who gave the explanation, but it came quietly and with a sense of discomfort. Baldip had begun drinking heavily, and seldom worked or took part in village affairs.

Once I had analyzed the land records systematically, the magnitude of the effects of this withdrawal became clear. He still held about thirty-two *bīghās* (most of which was actually in his mother's name), but as of the date of the record was obtaining the income equivalent to only eighteen. He had upgraded his canal land to well-canal and was farming that himself, but had let most of his well land on mortgage.

The discipline shown, in quite different ways, by Babu Ram, Sher Singh, and Indar Singh is not unusual. Nor, on the other hand, is Baldip Singh's abandonment of it. Resources are still scarce and the system is intensely competitive. Everyone knows the systems described here; every reflective person knows they are separate, so no one can look for one simple way to formulate the strategy that will allow his (or her) family to move ahead, or even hold its place, in the community. The strain of working and concentrating to balance all the various requirements of the resources one uses is sometimes too much to bear without respite, and those who break under it often do so fairly dramatically. In 1965 the village had its complement of madmen and madwomen. There were only a few of each, but the number is less important than the fact that they were generally described as having been broken by strain, misfortune, and being unable to enforce their demands for respect from others. In 1978 the madmen and madwomen were still there, but in addition frequent excessive drinking had become much more common as a middle ground between working at full concentration and giving up entirely, especially among men of about Netar's age and position—with enough money and time to drink but not enough responsibilities or prospects to see much danger in it. Nevertheless the basic pattern had not changed: those who could sustain the concentration inevitably set the competitive standard, for themselves and for others.

CONCLUSION

Social scientists sometimes suggest that social organization, or social structure, is nothing more than the sum of individual actions in a community. This confuses strategies with message sources and is precisely like saying the rules of chess are nothing more than the sum of all moves in all games, or the grammar and lexicon of a language are the sum of all utterances. This is nonsense—and dangerous nonsense because if one believes it, one could easily take the inherent fluidity and dynamism of strategies as evidence either that there is no structure at all or that structure is only in the eye of the beholder. There are rules and there are games. There are tools and there are tasks. In each case one should not be confused with the other, one cannot be reduced to the other, and neither can be ignored if you want to understand what sort of people the game players and tool users are—or if you plan to help them. Message sources provide the resources of strategies, the very stuff of decisions or action, people formulate strategies that bring the message sources to life and ultimately lead to their alteration or stability.

In Shahidpur between 1965 and 1978, all the strategies changed in the sense that the elements used and the players all changed, but this is the kind of change that is part of life at any time. In a more basic sense, the strategies didn't change at all. The discipline and concentration they required were the same as always and presented the same fundamental problems of judgment and action to the users.

CHAPTER 10

Cultural Systems, Rationality, and Change

The Green Revolution in Panjab was a very complex set of events. The changes were ecological, managerial, economic, religious, and political. There was no one policy that led to it, no one invention, and no one institution. Innovations were important, but so were continuities. The innovations had many sources. The Panjab government (through various rather autonomous agencies) was responsible for some, the central government for some, and private individuals and institutions for others. Some initiatives of the governments were crucial, but other important government actions consisted simply in removing prior regulations and practices that had held development back. It was important that the people very quickly accepted certain governmental innovations and it was equally crucial that they were able to resist others. All this means that instead of searching for the one cause of change, we have to search for the common ground in the many causes.

For those changes initiated by villagers and private individuals, the common ground is obvious enough: the aggressive and intelligent search for ways to get ahead in a highly competitive situation without disrupting the order that everyone depended on. For the government, the common ground is that the measures that were effective were those based on respect for the villagers' good judgment. The ineffective measures ranged from those that the villagers considered innocently ill conceived, to those backed by "arrogance" as their sponsors refused to accept the villagers' rejection as rational and tried to force acceptance by coercion or destructive inducements. This is not simply my private opinion—precisely the same thing was said by many in government.

Since I have argued that villagers' strategies were predominantly rational and efficient and that the arrangement of information in the message sources they use is likewise, as a legacy of successful strategies, rational, efficient, and logical, it is important to say just what this "rationality" is.

In characterizing rationality it is necessary to avoid identifying it with the content of any particular sets of rules or procedures, such as the rules of symbolic logic (cf. the discussion in Wilson, 1970) or the rules of micro-economic analysis and Western bookkeeping. Any rule or procedure can be used stupidly; none has ever been shown to be appropriate in all circumstances. Rationality lies not in what systems of ideas are used, but in the way they are used.

RATIONALITY

For the better part of this century, Western economic writers (among others) have generally represented peasants as irrational. Although the term "irrationality" was itself seldom used and proofs were never direct, accounts consistently held that peasants were unscientific, tradition-bound, liable to go seriously into debt for purely social or ceremonial reasons, inefficient in the use of labor and other resources, and generally inclined to operate their farms in such a way that, when costs were properly assigned and assessed, it would be found that they would have been better off either farming very differently or not farming at all.

In the early 1960s Theodore Schultz raised one of the few voices against this conception. Although not quite addressing the issue of rationality itself, he compared a village in Guatemala and one in India and argued that there were no significant "indivisibilities" of production in either (Schultz 1964:10ff.). This meant that every opportunity to exploit each resource at its maximum level of productiveness had been utilized. The peasants were "poor but efficient" (pp. 46, 49). Schultz drew on a study of a village in north India by David Hopper and an earlier anthropological study of a Guatemalan village by Sol Tax (1953). Tax had made precisely the same point. Indeed it was implicit in his title: *Penny Capitalism*. In 1968 Michael Moerman's *Agricultural Change and Peasant Choice in a Thai Village* focused more directly on the problem of rationality and its empirical description, and at the same time broadened the scope of the argument beyond economic choice and economic efficiency alone. Moerman argued that economic and technological choices were made in the context of general cultural values, which were separate from them, but that overall the peasants nevertheless acted like "cultural sharpies" rather than cultural automatons.

Due perhaps as much to the success of the peasants themselves as to the quality of these early arguments, Schultz has recently become the center of a very influential body of development-oriented economists, and in anthropology and other social sciences one hears more and more about "rationality hypotheses," "adaptation," and manipulation of cultural resources. So much so that the latest round of arguments seems to be dominated by Schultz's recent contention that:

> In terms of planning for economic development . . . I listed . . . a number of topics that should be placed on the back burner so we can get on with those awaiting analysis. Among the problems that should be placed aside are the following: Whether or not farmers respond to changes in price; the extent to which the existing agricultural resources are allocated efficiently under conditions of traditional agriculture; . . . the extent of surplus farm labor under these conditions. (Schultz 1968:356)

This adds up to saying that we can now assume rationality; we do not need to spend more time trying to prove its existence.

Kusum Nair's *In Defense of the Irrational Peasant* (1979) goes still further. The book is primarily an attack on the concept of rationality and its abuses in planning, as if it were indeed the dominant idea. Specifically, she seems to say that it is pointless to try to develop agriculture by providing price incentives and eliminating disincentives; such approaches have not worked and cannot work. Part of the reason, apparently, is that they do not take into account differences in character—by which she seems to mean differences between "communities." A Rajput, other things equal, will always be a Rajput, and a Jat will be a Jat. And the Jat, she seems to say (quoting the Jats she talked to without disagreement), will be the better farmer.

I would not in any way want to dispute Schultz's basic argument for the overriding importance of concentrating development efforts on solving specific technical, technological, and educational problems. But it doesn't follow that the best way to do it is by stopping pursuit of the general idea of rationality before we give it a clear formulation.

Nair is undoubtedly correct in saying that "rationality," as the idea is presently used, is subject to abuses. She is also right in pointing out that the same measures that created the successful "Green Revolution" in Panjab have not done so in Bihar. I am inclined to think that this is not because they were based on assumptions of rationality, but rather that the planners' understanding of rationality was insufficiently well worked out. In 1962, before any of this was envisioned, T. S. Epstein published her comparison of the reception of some related improved technology in two villages in south India. She showed in detail how "the same" government policies met two quite different responses—but for reasons that were equally cogent in both cases. This is reason enough to push on to the point where the general issues of rationality not only come back together with Schultz's practical recommendations but also support and provide a guide for them. In fact, we are within a finger's breadth of that point already.

If Babu Ram, Sher Singh, Gurdev Singh, Indar Singh, and Netar Singh, among many others, are not rational, then nobody is. Their major choices and actions always have, and are assumed by others to have, explanations. They have reasons for what they do, and these reasons are such that when others understand them, they can see how they would have been constrained to do the same things under the same circumstances. Thus the question of rationality boils down to this: what gives their acts this orderly, appropriate quality? What makes their decisions compelling?

ASPECTS OF RATIONALITY

There is no "essence" of rationality. Rationality, whether it is economic rationality, ecological rationality, managerial rationality, or rationality in terms of ideas of kinship, religion, or politics, has three major attributes which define a system of thinking or communicating that is independent of the different substantive contents of the several systems. They are consistency, relative efficacy, and accurate anticipation of the reactions of others.

Consistency

Consistency means that the behavior has to preserve the basic integrity of the cultural system in terms of which it is constructed. The basic learned definitions must be used consistently and as they are learned. It is rational to get good wheat seed in order to grow good wheat, to sell for more than costs to make a profit, or to want to understand the basic teachings of the guru to be a Sikh. It is irrational to plant barley and expect it to produce good wheat. It is irrational to buy seed for Rs. 30, invest Rs. 200 in making it grow, and expect a profit by selling the result for Rs. 150. It is irrational to claim that one is a good Sikh because one is a Jat, and Jats are better than other *jātīs*. Rationality is, in part, using one's cultural resources in ways that do not destroy the internal order that the use itself depends on.

Efficacy

Over and above simple consistency, there is also a sense in which rationality involves getting ahead. This is of course a matter of degree. Economists who have considered efficacy quite thoroughly in the context of the notion of economic rationality have been consistent in pointing out that there is much in such getting ahead that cannot be attributed to a quality of mind or an actual process of thought. Since decisions are undertaken in competitive settings, there can be much that is accidental in determining which of a series of equally consistent actions is in the long run the most efficacious. It is entirely possible that efficacious behavior will be undertaken for conscious reasons that are either not rational in the first sense at all, or not even related to the outcome. Such considerations have led economists to believe they could not and should not attempt to provide a characterization of rationality as an empirical phenomenon at all. Nevertheless, there is no reason to say that because we can in a strict sense speak of rationality in degree and connect it to competitive success, we therefore have to say that all actions which lead to someone getting ahead are rational. It is equally logical and much more useful to say that rationality is only one way to get ahead. We can go on to extract the important point: that there may be multiple ways to be consistent, but some will produce more-competitive advantages than others. The next question is how they can do so.

How can a person be sure that his consistent acts will also be efficacious? This has two answers. The first is an extension of the idea of following rules: one must also attempt to optimize the rules or choices in several systems of rules (such as the ecology and the economy) together, thereby minimizing the losses that the choices in one frame of reference force one to make in another. This involves being able to interpret the rules practically, to avoid being preoccupied with what people say, and to inform one's interpretation by observing the objects toward which the ideas direct attention. Like Sher Singh seeking his new farm, Indar with his constant experiments, or the participants in the debate over the proper basis of the Panjabi Subah, one has constantly to explore the limits and assess the force of the ideas one uses, even while using them. The second answer is that once this is done one must also look to the reactions of others.

Anticipating Reactions

The way we anticipate the reactions of others in applying cultural categories was first suggested by David Hume in explaining why we can consider a certain act evil under some circumstances but may not consider exactly the same act evil under different ones. His view was that when we make such a judgment about the quality or meaning of an act, we are at the very same time making a recommendation to others. The underlying idea is that judgments are also communications, which is directly related to my own argument that the elements of our cultural systems of ideas are at one and the same time bases of decision and assessment and bases of communication. Rational (or successful and rational) acts are, fundamentally, not only acts which are consistent in their basic definitions and optimal with respect to other decisions made by the same decision maker, but also are consistent in this larger context as well: in anticipation of the way others will react to these uses of the common cultural ideas and resources. Making rational decisions, behaving rationally, in this sense can be a complicated business.

Assuming that an action being contemplated is internally consistent (growing wheat from wheat seeds; being a Sikh while rejecting *jātī* discrimination), there are still three basic ways others can react and three basic effects those reactions can have on the outcome of one's own decision. The reactions of others can be to replicate or repeat the action; to do something recognizable as contrasting with it, logically related but not the same; or to do something that is unrelated (or not do anything recognized as related). In the two first cases, the response will be defined by the same context; in the last case the context will be different. The effects of any of these types of responses can be supportive, opposed, or neutral.

There is no inherent relationship between the reactions and their effects on the results of one's own choices. Each case has to be judged on its merits in the context of every system of ideas that can apply to it. For example, the decision to get a tractor will be supported, in an economic sense, by other peoples' decisions not to do so. The more people have tractors, the less favorable the situation will be for each—at least after a certain point. Not only will it be more difficult to hire out one's services for contract work, but eventually a large number of tractors can be logically expected to create rising fuel costs, fuel shortages, and probably a declining market for used tractors. From a management standpoint, if tractors become truly common, it is possible that theft may also become practical and security problems could increase. Nevertheless, in 1978 it was obviously felt that the point of diminishing value was not even close. According to a letter from Surindar in the summer of 1980, the number of tractors in the village had increased to ten and he hoped that it would increase still further.

By contrast the decision on the part of an individual to educate his children is not only supported by the fact that others make the same decision, it is impossible to carry out *unless* others do. Will the value of an education be diminished by the fact that many others also have one? Some people (but not villagers) say yes, and point to numerous B.A.s in India without jobs. From the village point of view, the assessments of education differ in the context of each major message source, but all nevertheless point in the same way.

Ecologically, the most spectacular effect of education is of course the creation of new plant varieties and technology, without which nothing else would be possible. For those who do not go on to become research scientists, there is still great scope for employing new technical knowledge, and the ability to get knowledge, at almost all levels of farming and production. A group of what the University Extension Service calls "progressive farmers," with fairly substantial holdings around the state, have made a tremendous contribution to the growth of agriculture by adopting new varieties on an experimental basis under a wide range of conditions and on a scale far beyond what the University resources alone could provide, and systematically reporting back their results in semiannual meetings with University staff as the seasonal "package program" recommendations are drawn up. Even for the most modest farmers it is increasingly important to be able to read, to do basic mathematics, and to understand some chemistry and biology—to calculate well pump requirements, for example, or insecticide doses. As more and more people acquire these skills and use them this way, the skills will become progressively more, not less, essential.

Managerially, there are many aspects to the value of an education. One is that for farming families, as noted, nonagricultural income is extremely useful as a complement to income in kind from the land. Another is that as more and more equipment is owned, it becomes more important to be able to read and understand information on how to maintain and repair it. Both these points can be expected to become even more relevant as more and more families branch out in this way. At the same time, with less need for labor as a direct means of holding land, there is less risk in sending family members off to school. This too can be expected to be increasingly true as more people become ever better educated.

Economically, education has modest cost and great potential returns. Its cost for landowning families is low partly because of the managerial situation and partly because cash tuitions are actually quite low. It has low cost for poorer families, if they are of scheduled castes, because their tuition rates are even lower, although the income lost by keeping a boy of later high school age in school (for example) is a relatively higher portion of total family income than it would be for a landowning family. The gains for landowning families come again both from the importance of cash for leverage in holding land, and from the possibility of obtaining enough money eventually to move out of farming and into something entirely different, if it is more remunerative. For nonfarming families, the gains consist directly in the possibility of cash income on a full twelve-month basis instead of on the less-steady basis of agricultural labor, especially in view of the job reservations. All in all, every one of these situations is expected to apply even more strongly as more and more people are educated and more jobs calling for education are thereby generated.

From the point of view of the Sikh religion, education as presently offered in Panjab is a good thing in its own right, and becoming educated is not only good for the individual but a way of meeting the general obligation one has to seek truth and to work for the good of the community. More of it would be better, and would be expected to create even more respect and support for it.

From the point of view of kinship, educating one's children is a major way for parents to fulfill their obligations to provide a useful patrimony to their children, and to provide for their own security in old age. It is especially important as a way of helping make the young person appear valuable as a potential husband or wife, so as to command a good marriage. This too is only likely to increase.

Finally, politically, the value of basic literacy in Panjabi and English has been clearly established in fighting court disputes, and education at and above the Bachelor's level is becoming increasingly important among higher elected officers. This makes it important for villagers themselves to have more education in order to deal with such figures (if they do not become them). Education will clearly not, in the villagers' opinion, become less valuable in the foreseeable future. The villagers' request for a full-fledged higher secondary school was granted in 1979/80 (adding grades ten through twelve).

Any course of action, once begun, can be assessed in this way. After selecting representative individuals, look at their action in the light of their resources in terms of each information system they use, whatever such systems may be (and they may be quite different from those in Shahidpur). Then look at the likely responses of those who can affect the realization of expected benefits. Then finally look again at the original calculations in light of these responses, to see if a reassessment is required. This may seem like an impossibly open-ended and complex agenda for analysis in the abstract, but it is not that difficult in practice. This is precisely because the number of points for choice that emerge in the process of public communication are vastly fewer than what one might imagine on purely logical grounds, just in the way that what is settled on and established to be taught in a culture is of much smaller bulk than what is ultimately learned. Therefore each person is likely to consider only relatively few options out of the total that are logically possible and these choices are also the ones most likely to be seen by others as well. It is the combination of choices that becomes complex in the formation of strategies. But to predict the consequences of specific choices, it is not necessary to find a way to aggregate the life strategies of all those one deals with. Tactics, in a sense, still remain the largest unit of choice one has to deal with. To assess the consequences of buying a tractor, one looks primarily at the likely consequences of the way others will make this same choice. One doesn't try somehow to add up into a single sum the total lifetime careers of all who might make that choice among others.

DESIGNING INNOVATIONS: TWO RECOMMENDATIONS

Undoubtedly some will construe the statement that villagers are rational and village behavior is efficient as a claim that this is the best of all possible worlds and there is no need or scope for improvement. Such inferences follow only if one is inclined at the outset to blame the poor for their poverty: to assume that the problems can lie only at the local level. In reality, they seldom do. Much more often, they lie in the resources that are provided to villagers from larger social systems, systems that the villagers themselves do not and cannot control.

These problems may stem from inappropriate or defective technologies and require the development of wholly new alternatives. They may stem from lack of basic facilities, like electricity and roads, and require large societal investments. Very often, however, they stem only from unrecognized aspects of the organization of government and other extralocal institutions that prevent adequate cooperation and thereby impede delivery of old but still potentially productive technologies, or even of readily available basic resources, to the needed places at the needed times in the right amounts and/or on reasonable terms. Both the design of resources and the provision for their delivery are encompassed in what Schultz means by "technological" problems, and very often the identification and solution of such problems—a problem being anything that stands in the way of improved productivity and increased welfare—are remarkably simple once one takes a systematic perspective. Two such problems, with their solutions, became obvious during the present study, and they will readily illustrate the point. One possible improvement has already been mentioned: credit for house construction. The second is the provision of rural telephones.

At the present time, lack of credit for rural housing clearly reduces demand for what would be a highly productive type of capital investment throughout India.

Credit for rural housing presently would not fall clearly into the areas of responsibility assigned either to the central government or the state. The central government has housing projects in the cities, where its tax base is, and sponsors some projects for underprivileged groups in the villages, through the states, apparently on the authority of the "directive principles" in the Constitution. But since rural housing is village housing, and villages are for the most part the administrative purview of the states, this is not a clear area of central government responsibility, and central agencies do not have the means to act uniformly in the ways that would be necessary to support such a program—to install a consistent system of title records, for example. The state, on the other hand, has the administrative apparatus to create and maintain the necessary legal records in the villages, but it is constitutionally restricted to legislating only in regard to agriculture and to collecting taxes on agricultural land, agricultural income, and water. The Constitution says nothing about rural housing, yet all matters not specifically on the state list are left to the central government. Up to 1978, officials in the Panjab cooperative system had enough to do without worrying about providing housing credit and were content to consider housing nonproductive. Banks did not generally operate in villages, or provide such credit either, although in 1978 the former aspect of the situation was changing.

It is not hard to see that the present problem could be resolved in either administrative channel (although it would be more natural to assign it to the states) without a very great effort. When the necessary protections can be provided both to lenders and borrowers, it is easy to consider villagers and ask what their reasons might be for accepting or rejecting such credit, and to see (as already suggested) that they would have many reasons for accepting it and few for rejecting it. It is equally easy to see that the housing industry could be the largest and healthiest of all the offspring of the Green Revolution—creating jobs and improved welfare on a scale as great as that stemming from the increases of direct farm production.

Rural telephone communication also did not exist in 1978 and, so far as I could tell from talks with administrators and cooperative officials, had never even been considered. From my own viewpoint—looking back at the way telephones followed rural electrification in the United States and greatly improved the condition of rural life at extremely little cost (none, to the government)—the absence was quite striking and the usefulness obvious. There were roads over which trucks could come to pick up crops, or deliver bricks, or over which ambulances could come from nearby clinics. But in order to get the trucks or ambulances it was necessary to send someone on a bicycle (at the fastest). A public telephone, perhaps in each village panchayat office, would enable villagers to derive far more benefit from the improved transportation and communication facilities at an additional cost that could not even begin to approach the gains. When I suggested this in the village and explained the American experience, the idea was immediately accepted with enthusiasm and even more uses were suggested—among the most striking being that of calling the police.

Why were there no telephones? Administratively, telephone communication in India is part of the responsibility of the Posts and Telegraphs Department, which is purely under central government authority, and in consequence, again, telephones have naturally been made available in cities and towns. They were allocated to businesses and houses, and the fees were very substantial—far above what any single village householder might be expected to afford. There was little reason for the urban-based telephone authorities to think of villages, and hardly anyone in a position (literally) to do so. On the other hand the states, although in a sense mandated to think of villages, had no one in a position (again, literally) to think of telephones. The obvious course would be for the state to ask the central government to install the telephones not in individual houses but in public places, like panchayat offices, and let the villages finance them through the cooperatives, recovering the expenses by charges for calls or a small village subscription. The villagers thought this would work, and that no one would mind paying enough on a per call basis to make the scheme viable. When one expects rational reasons for the status quo and assumes rational responses to solutions, the expectation is seldom disappointed and the remedy often lies close at hand.

THE BASIS OF CHANGE

No other place in the world can have precisely the same mix of institutions, ideas, and resources as Panjab, but this does not mean that we cannot expect similar progress elsewhere. Rationality is the key. Nothing else can explain the way villagers brought the productive resources together into a coherent package and rejected the resources that were useless and the policies that might have been destructive. Recognition of rationality *as a matter of policy* on the part of the government was basic to government's own constructive role. In 1965 I met many people in the government, particularly at higher levels, who understood the villagers' basic ideas, and saw the consistency in their actions, understood the competitive situation in villages (although their information on specific strategies was often incomplete), and realized that villagers were much more like themselves than was usually portrayed. But their outlook did not predominate over that of the functionaries who were appropriate to a government still interested more in control than development.

In 1978 it seemed the balance had shifted; the attitude of respect was not something one found here and there, but had become an explicit working assumption in the major departments concerned with rural matters. Measures to assure responsiveness began at the top of the state government and ran through the entire hierarchy. At the highest level, the Chief Minister, representing the majority in the legislature, had been given the power to review and alter civil service appointments to the seniormost positions in the administration. At the district level, the Development Officer, second in authority only to the District Commissioner, had been given broad power to coordinate the activities of the several previously independent divisions—canals, electricity, cooperatives, health services, police, and the like. At the same time, he had been made responsive to the Panjab Agricultural University scientists and was bound to assure that their advice was transmitted to and faithfully carried out by the Block Development Officers and Village Level Workers under his supervision. At the block level, the panchayat samitis had become increasingly important. And in numerous key agencies, including the cooperatives and the P.A.U., steps had been taken to create institutional arrangements to receive and analyze village reactions to whatever they were doing.

The new attitude was the result of all these various arrangements, and also of the fact that they worked. Those who had been involved in the changes and spoke of them to me obviously found their work exciting and rewarding. Agencies and institutions that failed to develop such links simply were not part of the process and generally also were not prospering.

A person can see what is in his own interest, and he can see what is in the interest of the wider system of groups and relations around him (under any of the sets of ideas that define such wider systems). A bad policy is one that forces him to choose between an action he sees as necessary in his self-interest and an action that he would see as leading to the larger good. The gain in governmental efficacy and responsiveness between 1965 and 1978 can mainly be measured by the reduction of instances when government action made such a calculation necessary, and this leads to what I would offer as a general formula to promote development. Develop basic resources, of course. Develop new technologies, of course. Provide basic transportation and coordinate the delivery of interdependent inputs, of course. But to be sure you are indeed doing all of this, find out if the people who are supposed to benefit will accept what you offer as good on balance, without major rational reservations or regrets. Establish channels for response from those concerned and support them with sound and comprehensive social research to discover the situation that the responses reflect. Are the proposed improvements in the best interests of those who are expected to accept them and in the best interests of those they live among as well? Are they consistent with what is already depended upon? Will they be efficacious compared to other means available? Above all, if one person accepts the innovation, what will the response of others be, and is this going to better the whole community from the point of view of those within it? No person in a position of authority should assume that most others are incapable of judging what is good for them. Nor, at the next step, should he assume that most others are interested only in their personal gain while he alone is interested in or capable of making a judgment concerning the general welfare. It is the mark of harsh and ultimately ineffectual ideologies that they condone such self-deception. Society is indeed an ongoing creation of those who live in it and through it. It is as complex and many sided as they are. It cannot be understood or changed if this fact is ignored, if society is assumed to be something independent in its own right and if the people are assumed to be its creations. Of course there are stupid and mean people in every community and in every institution, but by and large they do not rule. Perhaps no one is completely rational, but the vast majority are rational enough to recognize those who set constructive rather than destructive examples, to follow them and thereby to make them authoritative. These are the people the analyst or politician should consult. By and large in Panjab this is what happened.

The Iśa Upanishad, speaking of God as the basis of all knowledge, says that it is "within all this, and outside all this; past every thing in the moving world, this goes standing." So it is with the basis of progress. It is both within each social institution and category of person, and outside them all. Beyond all the institutions and categories, this remains unchanged: human reason and human energy. It was not one human device or creation, but human creativeness itself that was unleashed to bring about the Green Revolution. Here in Panjab there has been one of those bright moments in human history when wisdom has come out ahead of foolishness; kindness and fellow feeling have come out ahead of selfishness and fear.

It remains for history to show whether these twelve years were the full circuit of a day leading to a new night, or the first glow of a long dawn. Despite the recent difficulties, I am sure, however, that the villagers of Shahidpur will say it is the latter. The bright sun is just touching the rooftops, the crops are still damp with dew, the people are rising fresh for their tasks, and the Song of Hope still echoes from the village walls and the distant hills.

Bibliography

ARTICLES, BOOKS, AND MONOGRAPHS

Agriculture Department, Panjab

1964 *Package Programme for Intensive Cultivation of Rice in the Panjab*. Chandigarh: Controller of Printing and Stationary.

Bailey, F. G.

1960 *Tribe, Caste and Nation*. Manchester: Manchester University Press.

Beidelman, T. O.

1959 *A Comparative Analysis of the Jajmani System*. Monograph No. 8. Ann Arbor: Association for Asian Studies.

Bennett, Robert L.

1967 "'Surplus' Labor and Agricultural Development—Facts and Theories." *American Economic Review* 57:194–202.

Beteille, Andre

1965 *Caste, Class, and Power*. Berkeley and Los Angeles: University of California Press.

Billings, Martin H., and Arjan Singh

1971 "The Effect of Technology on Farm Employment in India." *Development Digest* 9(1):98–107.

Breman, Jan

1974 *Patronage and Exploitation*. Berkeley and Los Angeles: University of California Press.

CARE
1974 *Nutrition in Panjab*. Chandigarh: Punjab Nutrition Development Project.

Codere, Helen
1968 "Money Exchange Systems and the Theory of Money." *Man* 3:557–577.

Dalton, George
1965 "Primitive Money." *American Anthropologist* 67:44–65.

Darling, Sir Malcolm
1947 *The Panjab Peasant in Prosperity and Debt*. 4th ed. Bombay: Oxford University Press.

Dumont, Louis
1972 *Homo Hierarchicus*. London: Granada Publishing.

Economic Advisor to the Government, Punjab
1974 *Statistical Abstract of Punjab, 1973*. Chandigarh: The Economic Advisor to the Government, Panjab.

Economic and Statistical Adviser to the Government, Punjab
1977 *Statistical Handbook of Punjab, 1976–77*. Chandigarh: Government Printing Office.

Epstein, T. S.
1962 *Economic and Social Change in South India*. Manchester: Manchester University Press.

Falcon, Walter P.
1971 "The Green Revolution: Generations of Problems." *American Journal of Agricultural Economy* 52:698–710.

Frankel, Francine
1971 *India's Green Revolution: Economic Gains and Political Costs*. Princeton: Princeton University Press.

Garfinkel, Harold
1967 *Studies in Ethnomethodology*. Englewood Cliffs, N.J.: Prentice-Hall.

Gill, M. S.
1976 "The Green Revolution—Success in the Indian Punjab." *Progressive Farming* (December):7–15.
1980 *Panjab Maintains Lead: A Review of Agricultural Production in Punjab, 1973–79*. Chandigarh: Department of Agriculture, Panjab.
1983 "The Development of Punjab Agriculture, 1977–1980." *Asian Survey* 23(7) (July 1983):830–844.

Gulati, Kailash Chander
1974 *The Akalis Past and Present*. New Delhi: Ashajanak Publications.

Hough, Eleanor M.

1966 *The Co-Operative Movement in India.* 5th ed. revised by K. Madhava Das. London: Oxford University Press.

Hutchison, John A.

1981 *Paths of Faith.* 3d ed. New York: McGraw Hill.

Inden, Ronald B., and Ralph Nicholas

1977 *Kinship in Bengali Culture.* Chicago: The University of Chicago Press.

India, Government of

1978 *Constitution of India. As Amended up to the Constitution (Forty-third Amendment) Act, 1977.* Lucknow: Eastern Book Company.

Kahlon, A. S.

1971 "The Green Revolution: Its Promise, Performance, and Prospects." *Agricultural Development in Developing Countries—Comparative Experiences,* 402–404. Bombay: Indian Society of Agricultural Economics.

Kessinger, Tom G.

1974 *Vilayatpur: 1848–1968.* Berkeley and Los Angeles: University of California Press.

Leaf, Murray J.

1971 "The Panjabi Kinship Terminology as a Semantic System." *American Anthropologist* 73(3):545–554.

1972 *Information and Behavior in a Sikh Village.* Berkeley and Los Angeles: University of California Press.

1976 "Economic Implications of the Language Issue: A Local View in Panjab." *The Journal of Commonwealth and Comparative Politics* 14(2):197–203.

1979 *Man, Mind, and Science: A History of Anthropology.* New York: Columbia University Press.

1981 "The Green Revolution in a Panjab Village." *Pacific Affairs Quarterly* 83(4):617–625.

1983 "Economic Development and Cultural Change in a Panjab Village." *Economic Development and Cultural Change* 31(2):227–270.

Leaf, Murray J., et al.

1974 "Rituals and Social Organization: Sikh Marriage Ceremonies." *Frontiers of Anthropology*, 123–162. New York: D. Van Nostrand.

Lewis, Oscar

1958 *Village Life in North India.* New York: Vintage Books.

MacAuliffe, Sir Max Arthur

1963 *The Sikh Religion.* 6 vols. Bombay: S. Chand and Co. First published in 1889.

Malaviya, H. D.

1954 *Land Reforms in India*. New Delhi: Economic and Political Research Department, All India Congress Committee.

Michel, Aloys A.

1967 *The Indus Rivers: A Study of the Effects of Partition*. New Haven and London: Yale University Press.

Moerman, Michael

1968 *Agricultural Change and Peasant Choice in a Thai Village*. Berkeley and Los Angeles: University of California Press.

Myrdal, Gunnar

1968 *Asian Drama: An Inquiry into the Poverty of Nations*. New York: Pantheon.

Nair, Kusum

1979 *In Defense of the Irrational Peasant*. Chicago and London: The University of Chicago Press.

Noss, John B.

1980 *Man's Religions*. 6th ed. New York: Macmillan Publishing Company.

Raj, K. N.

1969 "Some Questions Concerning Growth, Transformation and Planning of Agriculture in Developing Countries." *Journal of Development Planning* 1:15–38. New York: United Nations.

Rajan, S. V. Govinda, and H. G. Gopala Rao

1978 *Studies on Soils of India*. New Delhi: Vikas Publishing House.

Raman, Sunder

1977 *Fundamental Rights and the 42nd Constitutional Amendment*. Columbia, Mo.: South Asia Books.

Randhawa, M. S.

1975 *The Green Revolution in Panjab*. Ludhiana: Punjab Agricultural University Press.

Redfield, Robert

1955 "Societies and Cultures as Natural Systems." *Journal of the Royal Anthropological Institute of Great Britain and Ireland* 85:19–32.

Schneider, David M.

1965 "Some Muddles in the Models." *The Relevance of Models for Social Anthropology*. Association of Social Anthropologists, Monograph No. 1. New York: Taplinger.

1968 *American Kinship: A Cultural Account*. Englewood Cliffs, N.J.: Prentice-Hall.

Schneider, Harold K.

1974 *Economic Man: The Anthropology of Economics*. New York: The Free Press.

Schultz, Theodore W.

1964 *Transforming Traditional Agriculture*. New Haven: Yale University Press.

1968 "Production Opportunities in Asian Agriculture: An Economist's Agenda." Extract published in *Agricultural Development in Developing Countries—Comparative Experience (1971)*, 356–360. Bombay: Indian Society of Agricultural Economics.

Sen, Sudhir

1975 *The Green Revolution: Food and Jobs for All*. New Delhi: Tata/McGraw-Hill.

Shannon, C., and Warren Weaver

1964 *The Mathematical Theory of Information*. Urbana: University of Illinois Press. First published in 1949.

Sharma, Miriam

1978 *The Politics of Inequality*. Honolulu: University of Hawaii Press.

Singh, Jogendra

1968 *Sikh Ceremonies*. Chandigarh: Sikh Religious Book Society.

Singh, Kushwant

1963 *A History of the Sikhs*. 2 vols. Princeton: Princeton University Press.

Srinivas, M. N.

1976 *The Remembered Village*. Berkeley and Los Angeles: University of California Press.

Srinivasan, T. N.

1971 "The Green Revolution or the Wheat Revolution?" *Agricultural Development in Developing Countries—Comparative Experience*, 404–416. Bombay: Indian Society of Agricultural Economics.

Tax, Sol

1953 *Penny Capitalism: A Guatemalan Indian Economy*. Smithsonian Institution, Institution of Social Anthropology, Publication 16. Washington, D.C.: U.S. Government Printing Office.

Wharton, Clifton R., ed.

1969 *Subsistence Agriculture and Economic Development*. Chicago: Aldine.

Wilson, Bryan, ed.

1970 *Rationality*. New York: Harper and Row.

Winch, Peter

1958 *The Idea of a Social Science*. London: Routledge and Kegan Paul.

Wiser, W. H.

1936 *The Hindu Jajmani System*. Lucknow: Lucknow Publishing House.

Wittgenstein, Ludwig
1965 *Philosophical Investigations*. New York: Macmillan.
Wundt, Wilhelm
1916 *Elements of Folk Psychology*. Trans. by Eduard Leroy Schaub. London: George Allen and Unwin.

LEGISLATIVE ACTS

The Hindu Succession Act, 1956; Act No. XXX of 1956, with Notes and Explanations. Delhi: Sat Dev Verma (1956).
Punjab Land Reforms Act, 1972; Punjab Act No. 10 of 1973, as Amended and up to Date with Notes. Chandigarh: Jain Law Agency (1977).
Punjab Security of Land Tenures Act, 1953; with Notes. Chandigarh: Jain General House (1977).

COURT DECISIONS

D. G. Mahajan v. State of Maharashtra. *All India Reporter* 915–941 S.C. (1977).
Jagan Nath and Others v. The State of Punjab. 64 *The Punjab Law Reporter* 22–30 (1962).
Sucha Singh Bajwa v. The State of Punjab. 1 *Indian Law Reporter* 575–614 (1974).

Index